The Open University
S342
Science: a third level course

PHYSICAL CHEMISTRY
PRINCIPLES OF CHEMICAL CHANGE

BLOCK 6
SURFACE TECHNIQUES

THE S342 COURSE TEAM

CHAIR AND GENERAL EDITOR
Kiki Warr

AUTHORS
Keith Bolton (Block 8; Topic Study 3)
Angela Chapman (Block 4)
Eleanor Crabb (Block 5; Topic Study 2)
Charlie Harding (Block 6; Topic Study 2)
Clive McKee (Block 6)
Michael Mortimer (Blocks 2, 3 and 5)
Kiki Warr (Blocks 1, 4, 7 and 8; Topic Study 1)
Ruth Williams (Block 3)

Other authors whose previous S342 contribution has been of considerable value in the preparation of this Course

Lesley Smart (Block 6)
Peter Taylor (Blocks 3 and 4)
Dr J. M. West (University of Sheffield, Topic Study 3)

COURSE MANAGER
Mike Bullivant

EDITORS
Ian Nuttall
Dick Sharp

BBC
David Jackson
Ian Thomas

GRAPHIC DESIGN
Debbie Crouch (Designer)
Howard Taylor (Graphic Artist)
Andrew Whitehead (Graphic Artist)

COURSE READER
Dr Clive McKee

COURSE ASSESSORS
Professor P. G. Ashmore (original course)
Dr David Whan (revised course)

SECRETARIAL SUPPORT
Debbie Gingell (Course Secretary)
Jenny Burrage
Margaret Careford
Sally Eaton
Shirley Foster
Sue Hegarty

The Open University, Walton Hall, Milton Keynes, MK7 6AA

Copyright © 1996 The Open University. First published 1996. Reprinted 2002.

All rights reserved. No part of this publication may be reproduced, stored in a retrieval system or transmitted in any form or by any means, without written permission from the publisher or a licence from the Copyright Licensing Agency Limited. Details of such licences (for reprographic reproduction) may be obtained from the Copyright Licensing Agency Ltd of 90 Tottenham Court Road, London, W1P 9HE.

Edited, designed and typeset by The Open University.

Printed in the United Kingdom by Henry Ling Ltd, The Dorset Press, Dorchester DT1 1HD

ISBN 0 7492 51875

This text forms part of an Open University Third Level Course. If you would like a copy of Studying with The Open University, please write to the Central Enquiry Service, PO Box 200, The Open University, Walton Hall, Milton Keynes, MK7 6YZ. If you have not enrolled on the Course and would like to buy this or other Open University material, please write to Open University Educational Enterprises Ltd, 12 Cofferidge Close, Stony Stratford, Milton Keynes, MK11 1BY, United Kingdom.

CONTENTS

1	INTRODUCTION	5
2	SURFACE-SENSITIVE TECHNIQUES	6
	2.1 Summary of Section 2	7
3	CLEAN SURFACES	8
	3.1 Summary of Section 3	8
4	THE ELECTRONIC STRUCTURE OF SOLIDS AND SURFACES	9
	4.1 Summary of Section 4	10
5	PHOTOELECTRON SPECTROSCOPY (PES)	11
	5.1 General principles	11
	5.2 X-ray photoelectron spectroscopy (XPS)	14
	5.3 Ultraviolet photoelectron spectroscopy (UPS)	19
	5.4 Summary of Section 5	20
6	AUGER ELECTRON SPECTROSCOPY (AES)	21
	6.1 Experimental aspects	22
	6.2 Nomenclature in AES	22
	6.3 Summary of Section 6	24
7	THE STRUCTURE OF SURFACES	25
	7.1 Crystal structures of metals: crystal planes and crystal surfaces	25
	7.2 Notation for the structure of adsorbed layers on surfaces	33
	7.3 Fractional surface coverage	36
	7.4 Summary of Section 7	37
8	LOW-ENERGY ELECTRON DIFFRACTION (LEED)	39
	8.1 The LEED experiment	39
	8.2 Interpretation of LEED patterns	40
	8.3 Summary of Section 8	46
9	VIBRATIONAL SPECTROSCOPY	47
	9.1 Infrared Reflection Absorption Spectroscopy (IRAS)	48
	9.2 Electron Energy Loss Spectroscopy (EELS)	49
	9.3 Application of IRAS and EELS	51
	9.4 Summary of Section 9	52
10	X-RAY ABSORPTION SPECTROSCOPY	53
	10.1 Extended X-ray Absorption Fine Structure (EXAFS and SEXAFS)	54
	10.2 Summary of Section 10	59
11	MICROSCOPY: IMAGING ATOMS	59
	11.1 Summary of Section 11	62

APPENDIX 1 ELECTRON BINDING ENERGIES	64
OBJECTIVES FOR BLOCK 6	66
SAQ ANSWERS AND COMMENTS	68
ACKNOWLEDGEMENTS	77

1 INTRODUCTION

Surface chemistry is the basis of most of the chemical industry and is also of immense importance in biology, Earth science and environmental science. In spite of its importance, many developments of surface science date only from the 1960s. At that time the design of catalysts was largely empirical, based on the efficacy of known processes and on chemical intuition. Today, surface science is a frontier of current research. Catalyst design is underpinned by the application of dozens of surface-sensitive techniques that provide detailed images of surface atoms and information about their physical properties. In this Block we focus on several of the more important of these techniques.

First, though, it is useful to consider those properties that make atoms on the surface of a solid different from those in the bulk. Crystalline solids have structures in which the atoms pack in regular arrays. Iron, for example, has the body-centred cubic structure (*bcc*), as shown in Figure 1a. From the alternative ball-and-stick representation in Figure 1b, it is apparent that atoms on the surface are differentiated from those in the bulk by their coordination: within the iron crystal, each atom is surrounded by eight nearest neighbours, at the corners of a cube, whereas nearest-neighbour coordination at the surface is fourfold, as a consequence of the removal of the top layer of atoms.

Viewed as a surface, the top layer of iron atoms depicted in Figure 1a is perfectly flat, with the atoms lying in a plane, but real catalysts are generally structurally complex, being polycrystalline or even amorphous. To simplify matters from the structural point of view, surface science often uses single crystals, which, in principle, should have a uniform structure throughout the bulk and on the surface. Real materials, however, are never this perfect. At any temperature above 0 K, a nominally 'flat' surface will always have vacancies (missing atoms) and adatoms (additional atoms), and in very many cases 'flat' **terraces** are found to be separated by **steps**, one atom in height, and possibly containing 'kinks' (Figure 2). At each of these irregularities an atom has a different environment and so is expected to have different properties, especially chemical reactivity. Not surprisingly, the chemistry of surfaces is often diverse.

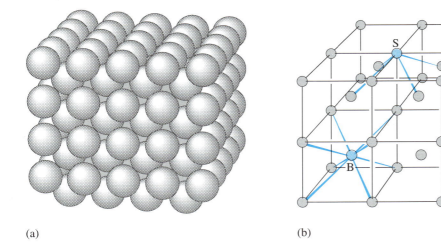

(a) (b)

Figure 1 The *bcc* structure of iron: (a) a space-filling model in which the atoms are represented by spheres in contact; (b) a ball-and-stick representation in which the atoms are represented by small balls in a lattice, denoted by thin lines. Links to *nearest* neighbours of the atoms S and B are shown in blue. The surface atom S has fourfold coordination. The bulk atom B has eightfold coordination.

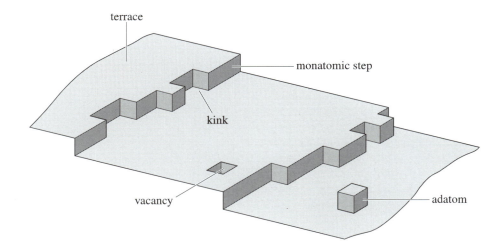

Figure 2 Schematic representation of imperfections on a surface. Each atom is represented here by a cube, so that a layer of atoms appears as a flat surface.

STUDY COMMENT Six of the Sections of Block 6 are devoted to techniques for studying surfaces. The emphasis throughout is on the information provided by each of these techniques, and you should check the Objectives and complete the SAQs to ensure that you can interpret results at the level required. Section 7 is about conventions in the representation of surface structures, and it provides the foundation for Section 8. Applications of some of the techniques are presented in video band 6 (*Electron spectroscopy*) on videocassette 2. You should view this video band immediately after your study of Sections 5 and 6.

2 SURFACE-SENSITIVE TECHNIQUES

A vast range of spectroscopic and other structural techniques has been applied to the study of solids. Some of these are applicable to surfaces. Indeed, some techniques are better suited to study surfaces than the solid bulk. The key to the design of surface-sensitive techniques is to avoid the observed signal being swamped by signals from the bulk of the solid, and this condition can be satisfied in various ways.

Structural techniques rely on the interaction of a probe with atoms of the solid, in order to generate a response. Probes are usually beams of electromagnetic radiation or of particles. Techniques that are surface sensitive require the beam to penetrate only the surface layers, *or* the resulting signal (response) to escape only from the surface layers. For example, irradiation of a solid by X-rays from an aluminium source in which the photon energy is 1 487 eV* results in the emission of electrons with energies in the range 0–1 482 eV. The photon energy is absorbed by an atom, from which an electron is emitted. X-rays penetrate deep into solids and are used in several techniques to investigate the structure of all types of material. On the other hand, electrons are strongly scattered by atoms. The mean free path of electrons, an average distance travelled in a solid, depends on their kinetic energy in the manner shown in Figure 3.

■ Approximately how many atomic layers are traversed by an electron with kinetic energy 100 eV? (Interatomic distances of closest neighbours are in the range 100–300 pm.)

□ From Figure 3 the mean free path of an electron of 100 eV is about 1 000 pm. So, typically less than ten atomic layers are traversed by the electron.

---

* eV denotes an electron volt, the kinetic energy gained by an electron that has been accelerated by a potential difference of 1 volt. 1 eV = 1.602×10^{-19} J, which is equivalent to 96 kJ mol^{-1}. We shall make much use of this unit in Block 6.

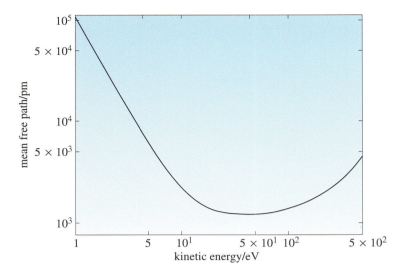

Figure 3 The mean free path of electrons in a solid material, such as a metal, as a function of kinetic energy measured in eV.

Electrons of suitable energy ($10-10^3$ eV) are therefore well suited to the study of surfaces, whether as probes (because they rarely penetrate beyond the first few surface layers), or as emitted signals such as photoelectrons (because their escape depth is limited to the surface layers).

Electromagnetic radiation of wavelength similar to interatomic distances (i.e. X-rays) is highly penetrating, and is consequently less directly suited to act *simultaneously* as both probe and signal in surface studies. One solution to this problem is to direct the probe beam at the sample at a glancing angle, so that it travels sufficiently far to have a high probability of interacting with an atom of the solid, yet at the same time remaining close to the surface. Most of the emitted signal can then emerge without further significant interaction with the bulk of the solid, as shown in Figure 4. Careful experimental design thus allows a range of electromagnetic radiation to be used directly and indirectly in the study of surfaces.

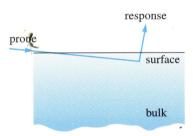

Figure 4 Probe–surface interaction at a glancing angle to the surface.

In practice, the majority of techniques used for studying surfaces use beams of electrons or photons. Electrons are strongly scattered and absorbed by all matter, including air. Several regions of the electromagnetic spectrum, especially the far ultraviolet and X-rays of relatively long wavelength, are also absorbed in air. For this and other reasons (see below) it is often necessary to study surfaces under conditions of very low pressure, and usually at ambient temperature, although much higher temperatures are sometimes used.

The motivation for surface studies, apart from the intrinsic curiosity of scientists, is to understand how catalysts work, especially those used in the chemical industry. Conditions in catalytic reactions are typically far from those used in laboratory surface studies; high pressures (commonly up to 100 atmospheres) and temperatures (up to 1 100 K) are often needed to give high yields. Moreover, laboratory studies are often made using the surfaces of single crystals, in contrast to the compacted powders used in commercial catalysts. In spite of these differences, surface science has led to a better understanding of many catalytic processes, and in recent years apparatus has been designed to enable laboratory studies to be made under conditions closer to those at which real catalysts operate.

2.1 Summary of Section 2

1   Surface techniques generally involve a probe and a response, either of which may be electromagnetic radiation or particles.

2   Electrons are especially suited as both probe and response, because they do not travel far in solids.

3 CLEAN SURFACES

In order to study the properties of the surface of a crystal, it is essential that the surface be clean; otherwise the experiment is likely to reveal more about the dirty covering than about the underlying surface. Clearly, the surface, once cleaned, must remain so during the course of the experiment. The conditions required can be estimated by considering a crystal surface of area 1 cm². Typically, this surface has about 10^{15} atoms in its top layer. At room temperature and pressure, molecules of air collide with this 1 cm² of surface at the rate of about 3×10^{23} per second. If all of these molecules were to stick to the surface, it would be completely covered with a monolayer in about 10^{-8} s or about 3×10^{-12} h. As the rate of coverage is directly proportional to the pressure of gas, the pressure must be reduced accordingly. So, if complete monolayer coverage is to be reached after only one hour, the time of a typical experiment, the pressure must be reduced by a factor of about 3×10^{-12}, that is, to 3×10^{-12} atm. But for a coverage of only a few per cent in an hour, the pressure should be further reduced to about 1×10^{-14} atm, which is about 1×10^{-11} torr*. Such pressures lie in the so-called ultra-high vacuum (UHV) region. Apparatus operating under these conditions is made of stainless steel because atmospheric helium diffuses through the walls of glass vessels, which prevents the pressure being decreased below about 10^{-10} torr.

Any substances put in such an apparatus will initially be covered with molecules adsorbed from the atmosphere. Several methods of varying degrees of severity are used to clean such a surface. Many physically adsorbed molecules will simply desorb at pressures as low at 10^{-10} torr, but where the molecules are chemisorbed, pumping at low pressure is ineffective.

- ■ How might chemically bound adsorbates be removed from a surface?
- ▫ Several methods exist, the simplest being heating, which breaks the adsorbate–surface chemical bond. Physical disruption of the surface is also effective (see below).

As it happens, many of the adsorbates that adhere most strongly are substances that may be readily oxidized, such as organic molecules. Heating in an atmosphere of oxygen, with subsequent pumping, is often employed. If the surface to be studied is itself easily oxidized, as many metals are, the unwanted oxide coating can be removed by heating in an atmosphere of hydrogen.

More severe methods of 'cleaning' the surface include etching and cleaving. Etching is usually done with a beam of high-energy (100–3 000 eV) argon ions, Ar^+. Argon atoms are massive enough to dislodge impurity species and even atoms of the parent material, leaving the surface clean but damaged, and, because it is an unreactive noble gas, it can easily be pumped away. The damage may then be rectified by heating at $T_m/2$ (where T_m is the melting temperature), a process known as annealing. Where single crystals are used, it is sometimes possible to cleave a thin slice from the surface by physically striking the side of the crystal, in a vacuum, with a chisel-like tool. In some cases a completely new and clean surface may be generated under UHV by evaporation from, for example, an electrically heated filament, onto the surface of a solid substrate, where a thin film of the evaporated material builds up.

3.1 Summary of Section 3

Surface techniques require the use of low pressure, often UHV, and usually also pre-cleaning.

---

* The non-SI unit of pressure, torr, is still used in surface chemistry, although bar and mbar are also used. 1 atm ≈ 1 bar = 760 torr; 1 torr = 133 Pa. The torr is the pressure that will support a column of mercury of height 1 mm.

4 THE ELECTRONIC STRUCTURE OF SOLIDS AND SURFACES

Many of the techniques used for studying surfaces and adsorbates examine the interaction of radiation with electrons bound in surface atoms. In this Section we revise some ideas about the electronic structure of solids and introduce a property called the work function, whose value often gives useful information about adsorbates. Among the various ways of classifying solids is the division into three groups — metals, semiconductors and insulators. In practice, the distinction between these groups is made according to their electrical conductivity, a quantity measured in the unit siemens metre^{-1} (S m^{-1}). **Metals** typically have a conductivity of more than 10^4 S m^{-1} and **insulators** less than 10^{-14} S m^{-1}. The conductivity of **semiconductors** lies between 10^{-8} and 10 S m^{-1}; the gaps in these ranges indicate the arbitrary nature of this classification. The understanding of electrical conductivity, and of this classification, depends on knowledge of the electronic structure of these types of material. Whereas in individual atoms and molecules the electrons occupy discrete and well-separated energy levels, in extended solids the discrete levels are so closely spaced that they are described as **bands**.

These broad principles were introduced in the Second Level Inorganic Course, and are illustrated in Figure 5. For electrons to be able to move through a solid and conduct electricity, they must have access to *vacant* levels of higher energy, into which they can be promoted when energy is supplied by the application of a potential difference across the material. In other words, electrons in a completely filled band cannot conduct. In insulators and semiconductors all the levels in the uppermost populated band, the *valence band* (VB), are fully occupied. At a higher energy, there is an empty band, the so-called *conduction band* (CB). The energy difference between the valence band and the conduction band is known as the *band gap*. It is the size of this gap which is the critical factor. In semiconductors it is small, typically about 1 eV, so that at ordinary temperatures *some* electrons are thermally promoted into the conduction band, where they do have access to vacant levels and so can carry a *small* current. In an insulator, however, the gap is much wider (> 4 eV), such that no electrons can reach the conduction band, and conductivity is zero (except at very high temperatures where insulator break-down may occur). Metals are characterized by an uppermost populated band (the conduction band) which is *partially* filled. In effect, the band gap has now disappeared, and the electrons in the lower part of the band have easy access to a large number of vacant levels in the upper part. Hence, conductivity is high.

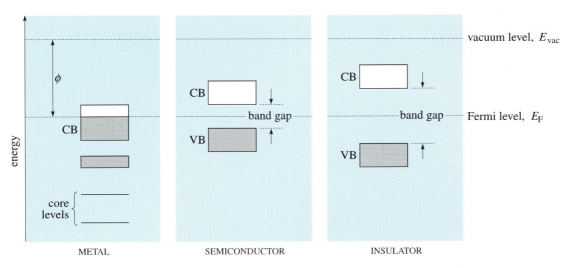

Figure 5 Schematic band description of the three types of solid: metal, semiconductor and insulator. Occupied levels are indicated by grey tint. The vacuum level, Fermi level and work function, ϕ, are all defined in the text (VB = valence band; CB = conduction band).

The **photoelectric effect**, the ejection of electrons from a solid by ultraviolet radiation, provides a convenient method for investigating band structure. As the photon energy is increased, no emission is observed until the energy is just sufficient to remove the least strongly bound electrons. At that point, excitation occurs from the uppermost occupied level in the solid to the **vacuum level**, E_{vac}, the level representing the free electron with no kinetic energy (Figure 5). In a *metal* the uppermost occupied level is known as the **Fermi level**, E_F. In *any* solid the difference in energy between the Fermi level and the vacuum level, $(E_{vac} - E_F)$, is called the **work function**, symbolized as ϕ.

■ How could the value of ϕ for a metal be measured experimentally?

□ The work function ϕ corresponds to the difference in energy $E_{vac} - E_F$. So ϕ is equal to the minimum photon energy that will cause photoemission from the surface of the metal, this being the difference between the energy of an electron at the Fermi level and that of a free electron with no kinetic energy. (For insulators and semiconductors, the Fermi level lies in the band gap (Figure 5), and so ϕ is less than the minimum energy for photoemission.)

As noted above, the work function of a metal surface, particularly one that is clean and flat, can be measured in a vacuum. For metals, ϕ has values between 2 and 6 eV, the exact value being characteristic of each metal and, for a given metal, characteristic of the particular plane forming the surface. Not surprisingly, the work function of a rough surface is different from that of a flat one, as various types of crystal plane are exposed in a rough sample (see Figure 2). Of more interest than the absolute values of ϕ are the *changes* in the work function, $\Delta\phi$, which occur as a result of adsorption. These can be either positive or negative and, although the interpretation is sometimes difficult, many examples can be understood in terms of charge transfer between the adsorbate and the underlying surface. Halogen atom adsorption on copper illustrates the point.

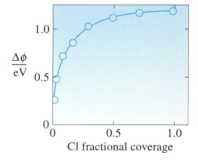

■ On the basis of a simple ionic model, how would you expect the work function to be affected by adsorption of chlorine atoms on a copper surface? [*Hint* The relevant Pauling electronegativities are $\chi(Cu) = 1.9$, $\chi(Cl) = 3.16$. The larger electronegativity of chlorine results in chemisorption in which the ionic character of the bond is in the sense $Cu^{\delta+}-Cl^{\delta-}$.]

□ With a larger positive charge on the copper surface, electron release is retarded and so the work function increases. In fact, for coverages as low as 10%, ϕ is increased by almost 1 eV from the clean-surface value of about 4.94 eV, as Figure 6 shows.

Figure 6 The variation of the change in work function, $\Delta\phi$, with fractional surface coverage, θ, of chlorine on the surface of a crystal of copper. The clean surface value is close to 4.94 eV.

From the intuitive notion of charge transfer in this example, we conclude that ϕ increases if the adsorbate gains negative charge *from* the surface, and that ϕ decreases if the adsorbate transfers negative charge *to* the surface. We also note from Figure 6, that the change in ϕ increases with surface coverage, tending towards some maximum value. In some cases, the work function is found to be approximately proportional to coverage. Measurement of $\Delta\phi$ then provides a very convenient means of estimating the fractional coverage experimentally.

4.1 Summary of Section 4

1 The electronic structure of solids can be described by band theory.

2 In band theory, metals, insulators and semiconductors are characterized by the relationship of the valence band and conduction band, and by the electronic occupancies of these bands. In a metal, the uppermost occupied level is known as the Fermi level; in semiconductors and insulators, the Fermi level lies in the band gap between the valence band and the conduction band.

3 For any type of solid, the difference in energy between the vacuum level and the Fermi level $(E_{vac} - E_F)$ is called the work function, ϕ. Changes in the work function on adsorption provide information about the nature of the binding of adsorbates to a surface.

SAQ 1 How would you expect an increase of temperature to affect the electrical conductivity of a semiconductor, according to the band description of its electronic structure?

SAQ 2 Figure 7 shows how the change in work function, $\Delta\phi$, varies with increasing coverage of a tungsten surface by lithium atoms. What can you deduce about the direction of charge transfer? Is this consistent with the Pauling electronegativity of lithium, which is significantly less than that of tungsten (χ(Li) = 0.98; χ(W) = 1.7)?

SAQ 3 Figure 8 shows the effect on the change in work function, $\Delta\phi$, of increasing coverage of a rhodium surface with carbon monoxide. What can you deduce about the direction of charge transfer? Is the variation of ϕ with coverage as expected?

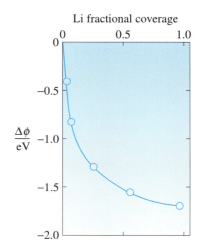

Figure 7 The dependence of the change in work function, $\Delta\phi$, on the fractional surface coverage, θ, of lithium on the surface of a tungsten crystal.

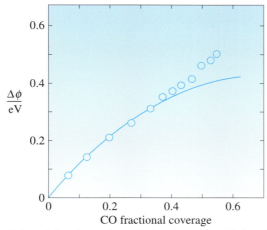

Figure 8 The variation of the change in work function, $\Delta\phi$, with fractional surface coverage of CO on the surface of a crystal of rhodium. (The circles represent experimental points; for an explanation of the plotted line see the answer to SAQ 3.)

5 PHOTOELECTRON SPECTROSCOPY (PES)

5.1 General principles

Photoelectron spectroscopy is the study of the photoelectrons emitted when a substance is exposed to monochromatic radiation — that is, to radiation of fixed frequency, v, and so fixed photon energy, hv.

The kinetic energy of a photoelectron, E_k, is related to I, the ionization energy of that electron in an atom or molecule of the sample, according to the **Einstein equation** (equation 1):

$$E_k = hv - I \qquad (1)$$

where hv = photon energy of the incident radiation, and I = ionization energy of the electron.

In PES the ionization energy is more commonly called the **binding energy**, which we represent by E_B. The relationship between E_k, hv and E_B (or I) is illustrated schematically in Figure 9.

The kinetic energies of these ejected electrons are analysed using, for example, a hemispherical analyser, which permits only electrons of particular energy to traverse the space between the two charged hemispheres shown in Figure 10. By changing the electric potential between the two hemispheres, electrons of various kinetic energies pass through the analyser, and so a spectrum of kinetic energies can be recorded.

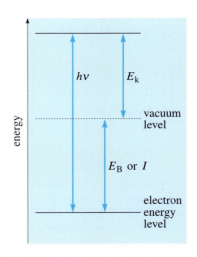

Figure 9 The relationship between the photon energy, hv, the binding energy, E_B (or ionization energy, I), and the electron kinetic energy, E_k.

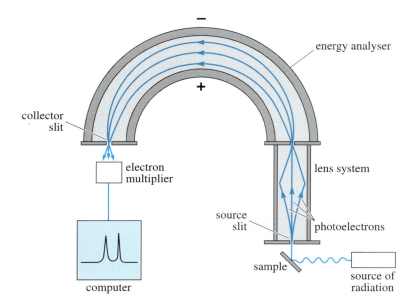

Figure 10 Cross-section through a photoelectron spectrometer with a hemispherical analyser of electron kinetic energies.

A photoelectron spectrum is therefore a plot of the intensity of the signal against the electron binding energy. The intensity of the signal (or 'photoelectron current') is proportional to the number of photoelectrons.

■ A schematic photoelectron spectrum of the noble gas krypton in shown in Figure 11. What do you deduce from the discrete values of binding energies, and how should the peaks in the Figure be labelled?

□ The discrete binding energies reflect the electronic structure (energy levels) of atomic krypton. The peaks may be labelled according to the values of the two quantum numbers n and l. In order of *decreasing* binding energy, from right to left in the spectrum, the peaks correspond to the subshells 2s, 2p, 3s, 3p, 3d, 4s and 4p. Note that the 1s level is inaccessible because $h\nu < E_B$ for the radiation commonly used in the PES experiment.

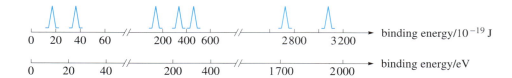

Figure 11 The photoelectron spectrum of krypton.

In answering this question you assumed implicitly that a relationship exists between the *experimental* quantity, binding energy, and the *theoretical* notion of an energy level. This relationship is expressed quantitatively by **Koopmans' theorem**, as stated in equation 2, which equates the *magnitude* of the binding energy with the orbital energy, ε. The orbital energy is the difference in energy of the electronic energy level (the orbital) and the vacuum level, and so has the opposite sign to E_B:

$$E_B = -\varepsilon \tag{2}$$

The principles of PES, applied to atomic krypton above, apply equally well to molecular substances. As shown in the Second Level Inorganic Course, the molecular orbital (m.o.) energy-level diagram for carbon monoxide is constructed from the atomic orbital diagrams of carbon and oxygen (Figure 12). Do not worry at this stage about the labels on this diagram, since we shall reconsider CO in Section 9.3. The important point to note is that the energy levels of electrons in molecules are quantized. Hence in the PES spectrum of CO we should expect to observe a distinct number of peaks, each corresponding to an occupied energy level in Figure 12. Indeed, this is what is observed; as shown in Figure 13, the PES spectrum contains one set of peaks for each of the occupied energy levels in the m.o. diagram. As this Block is concerned with the interpretation of experiments such as PES, you will not be expected to construct m.o. diagrams such as that in Figure 12.

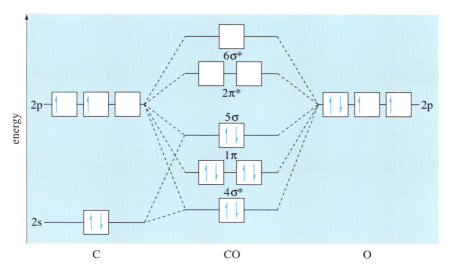

Figure 12 The molecular orbital energy-level diagram for CO.

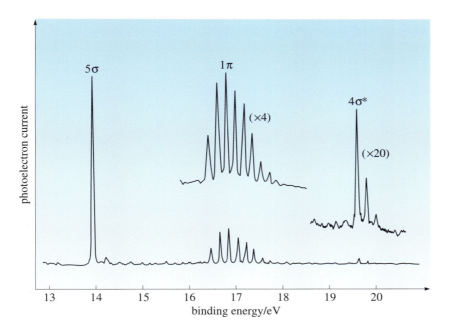

Figure 13 The photoelectron spectrum of CO. Three sets of signals are observed, at about 14, 17 and 20 eV, corresponding to the three sets of occupied molecular orbitals represented in Figure 12. Each set of signals consists of several peaks, resulting from vibrational energy changes; however, this does not concern us here.

The binding energy scales in Figures 11 and 13 reveal an important experimental aspect of photoelectron spectroscopy. Photon energies up to about 50 eV are in the ultraviolet region of the electromagnetic spectrum and are achievable using gas discharge lamps, similar in principle to fluorescent and street lights. Photons of several thousand eV in energy are X-rays and can be generated using X-ray tubes. Both types of radiation source, uv lamps and X-ray tubes, are used in PES. They form the basis of two distinct techniques, **ultraviolet photoelectron spectroscopy**, **UPS**, and **X-ray photoelectron spectroscopy**, **XPS**.

■   Another fundamental difference exists between UPS and XPS. Using the labels of the peaks in Figures 11 and 13, identify the basis of this distinction.

□   The photon energy of X-rays is so high that irradiation results in ionization of both the valence electrons and the very tightly bound electrons, the *core electrons*. The lower energy of uv photons makes only the *valence electrons* accessible to study by UPS.

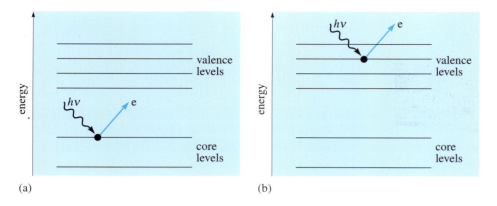

Figure 14 The photoionization processes for: (a) core electrons in XPS; (b) valence electrons in UPS (not to scale).

The processes involved in XPS and UPS are illustrated schematically in Figure 14.

The distinction in technique and energy levels between the two methods ensures that each provides a different sort of information, so that XPS and UPS are complementary techniques. UPS is limited to the study of valence electrons, for which it gives detailed information. XPS extends to core electrons and is not much used for valence electron studies. One reason for this is the inherent resolution of the two techniques, imposed by the characteristics of the ionizing radiation. X-ray sources as used in XPS typically have line-widths of about 1 eV, whereas uv (He lamp) sources have line-widths of about 10 meV.

5.2 X-ray photoelectron spectroscopy (XPS)

In XPS the source of X-rays in the laboratory is an X-ray tube. This is a tube in which electrons of high energy are fired at a metal target, which then emits X-rays. The metals most commonly used in XPS are magnesium and aluminium. For both metals a spectrum of X-ray lines is produced, but in each case it is dominated by an intense line, the K_α emission*, of photon energy 1 254 eV for magnesium and 1 487 eV for aluminium.

- ■ Using an aluminium X-ray tube, the XPS spectrum of a sample of steel is found to have intense photoemission of electrons with kinetic energies of 955 eV and 1 203 eV. What are the binding energies of the emitted electrons?

- ■ From equation 1, the binding energy is the photon energy minus the kinetic energy, which gives values of 532 eV and 284 eV.

These two binding energies are known to be characteristic of oxygen and carbon, respectively. As explained in Section 2, electrons of energy up to about 1 200 eV typically traverse less than ten atomic layers. So XPS is essentially a technique for studying surfaces rather than the bulk properties of a substance.

In application, XPS concentrates on the study of core electrons. As these are not generally involved in chemical bonding, we might expect that the binding energies of core electrons are relatively insensitive to the chemical environment of an atom. This is indeed borne out by observation. For example, the 1s electrons of carbon have binding energies in the range 280 to 305 eV, and in most materials the value is close to 284 eV. In fact, the values of binding energies of core electrons are sufficiently characteristic that XPS is a method of *qualitative* analysis. As all elements beyond the first row of the Periodic Table (that is, beyond hydrogen and helium) have core electrons, XPS is a universal method. Some typical values of binding energies for the first 33 elements in the Periodic Table are given in Table 1; a full list of binding energies is given in Appendix 1.

---

* K_α simply denotes the X-ray photon emitted when an electron makes the transition from the 2s subshell to the 1s (or K) subshell, where a hole has been created by the high-energy electron beam.

Table 1 Typical values of binding energies/eV[a]

	K	L_1	L_2	L_3	M_1	M_2	M_3
1 H	14						
2 He	25						
3 Li	55						
4 Be	111						
5 B	188						
6 C	284						
7 N	399						
8 O	532	24					
9 F	686	31					
10 Ne	867	45					
11 Na	1 072	63		31			
12 Mg	1 305	89		52			
13 Al	1 560	118	74	73			
14 Si	1 839	149	100	99			
15 P	2 149	189	136	135			
16 S	2 472	229	165	164	16		
17 Cl	2 823	270	202	200	18		
18 Ar	3 203	320	247	245	25		
19 K		377	297	294	34	18	
20 Ca		438	350	347	44	26	
21 Sc		500	407	402	54	32	
22 Ti		564	461	455	59	34	
23 V		628	520	513	66	38	
24 Cr		695	584	575	74	43	
25 Mn		769	652	641	84	49	
26 Fe		846	723	710	95	56	
27 Co		926	794	779	101	60	
28 Ni		1 008	872	855	112	68	
29 Cu		1 096	951	931	120	74	
30 Zn		1 194	1 044	1 021	137	87	
31 Ga		1 298	1 143	1 116	158	107	103
32 Ge		1 413	1 249	1 217	181	129	122
33 As		1 527	1 359	1 323	204	147	141

[a] The labels K, L_1, L_2, etc., indicate the energy level from which the photoelectron is emitted: K ≡ 1s, L_1 ≡ 2s, L_2 and L_3 ≡ 2p, M_1 ≡ 3s, M_2 and M_3 ≡ 3p. Their significance will be explained in Section 6.2.

■ Using the typical values of binding energies in Table 1, identify as far as possible the elements that cause the peaks labelled W, X, Y, Z observed in the XPS spectrum of a sample of aluminium shown in Figure 15. (Typically, XPS spectra give binding energies to about ±1 eV, although in reading the data in Figure 15 you can probably estimate the values to only about ±5 eV.) Take the values to be approximately 70, 115, 290 and 530 eV, and assume that no peaks occur at higher binding energies.

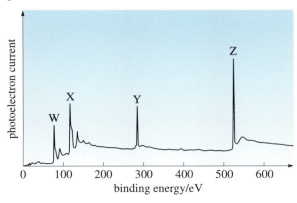

Figure 15 The XPS spectrum of aluminium.

■ From Table 1 the peaks correspond, in order, to the following ionizations: W = Al(2p), X = Al(2s), Y = C(1s) and Z = O(1s). You may have found it difficult to assign without ambiguity the peaks at E_B = 70 eV and E_B = 115 eV. The peak at E_B = 70 eV (W) could be caused by Ni(3p) or Cu(3p) ionization (68 and 74 eV, respectively); the peak at E_B = 115 eV (X) might equally be attributed to the Be(1s) signal, expected at about 111 eV, or the Ni(3s) signal at about 112 eV. In the interpretation of XPS, such ambiguities often arise. They are usually resolved by the presence or absence of other peaks due to the same element and by previous chemical knowledge of the sample. As in our spectrum there are no signals at E_B > 530 eV, it is possible to rule out Ni(3p) and Cu(3p) for the 70 eV signal, and to eliminate Ni(3s) for the 115 eV signal.

The minor peaks in Figure 15 would give further information on the composition and assist with the confirmatory evidence. The wide applicability of XPS as an analytical method is the reason for its alternative (now less widely used) name ESCA, from Electron Spectroscopy for Chemical Analysis.

Although XPS is a simple and excellent technique for identifying the elements present in a surface, it is less well suited to quantitative analysis. The intensity of a PES signal is dependent on a number of factors; in addition to the concentration of the element in question, the chemical environment and physical state of the sample, and the orbital from which photoemission occurs, all affect the intensity. Nevertheless, it is often possible to use XPS to make accurate *estimates* of the composition of the surface of a solid, and also to deduce an approximate depth profile of an element — that is, the change in composition with depth from the surface (although we shall not explore this aspect further here).

5.2.1 Spin–orbit coupling

The implication of the previous Section is that in a photoelectron spectrum a single peak is observed for each electron subshell. This, however, is an oversimplification. Figure 16 shows the PES spectrum of a sample of metallic zirconium (Zr). Whereas only one peak is observed for the 3s subshell, two peaks are observed for the 3p subshells. This doubling of the p peaks is due to **spin–orbit coupling**, an interaction between the spin of the electron and its orbital motion, as designated by the quantum number l. In terms of a classical picture, the electron orbiting a nucleus may be described equally as a nucleus orbiting the electron. Such an electric charge circulating around the electron creates a magnetic field, within which the electron may align parallel to the field or opposed to it. This results in two possible non-degenerate energies for the electron. The non-degenerate states are represented by a quantum number, j, which is given by the vector sum* of the orbital quantum number, l, and the spin quantum number, s. This quantum number has the values:

$$j = l \pm s \tag{3}$$

For a single electron, such as we are concerned with in the PES experiment, where each photon ionizes *one* electron, $s = \pm\frac{1}{2}$, so that j has the values $l \pm \frac{1}{2}$.

Take the example of an electron in a p orbital, $l = 1$. So the values of j are $j = 1 + \frac{1}{2}$ and $j = 1 - \frac{1}{2}$ — that is, $\frac{3}{2}$ and $\frac{1}{2}$. Notice that these are the subscripts of the labels attached to the two 3p peaks in Figure 16.

■ What values does j have for an electron in a 3d orbital ($l = 2$)?

■ Again $s = \pm\frac{1}{2}$, and so j has the values $(2 + \frac{1}{2})$ and $(2 - \frac{1}{2})$ — that is, 5/2 and 3/2, as the labels for the 3d peaks in Figure 16 show.

---

* The term *vector* means that the quantity has both magnitude and direction.

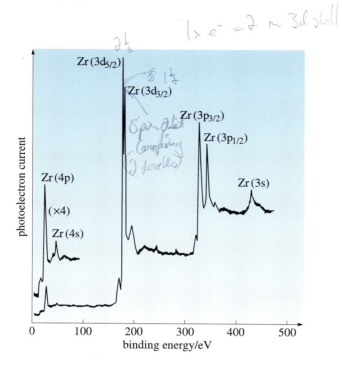

Figure 16 The PES spectrum of a sample of zirconium. The labels used in this Figure are explained in the text.

An electron in an s orbital has $l = 0$, and so there is no orbital contribution to couple with the spin, resulting in single s peaks in PES.

■ Look at the values of binding energies in Appendix 1. How does the effect of spin–orbit coupling change with atomic number (Z)?

▪ The effect is quite small for the light elements (low Z) and large for heavy elements, increasing with increasing values of Z.

This increase of spin–orbit coupling is quite consistent with the physical picture by which it arises, because the magnetic field created by the circulating nucleus will increase as the nuclear charge increases.

5.2.2 The chemical shift

Although the binding energies of electrons are sufficiently characteristic for XPS to be the analytical technique of choice for most surface investigations, the precise value of the binding energy for a given type of atom depends on its chemical environment. This variation of E_B is called the **chemical shift effect**. It originates from the interaction between the core electrons under investigation and the valence electrons. The effect of this interaction is readily understood through the example of the fluorine derivatives of methane, albeit in the gas phase. Recall that fluorine has a very high electronegativity relative to carbon.

■ How will increasing fluorine substitution in CH_4 affect the electron density at the carbon atom, and will this increase or decrease the C(1s) binding energy?

▪ An atom with high electronegativity withdraws electron density from carbon, leaving it with a net positive charge. As a result, all the electrons of the carbon atom are held more tightly and so the binding energies increase.

This prediction is indeed borne out, as Figure 17 shows. As the degree of fluorine substitution increases, the C(1s) binding energy (and hence the chemical shift) increases, being about 12 eV higher for CF_4 than for CH_4. Moreover, a knowledge of electronegativities enables the effect to be calculated, in the case of gases, giving good agreement with experiment, as the data in Figure 17 show.

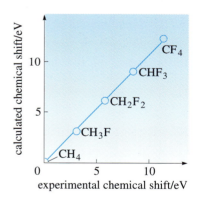

Figure 17 Correlation between calculated chemical shifts for the 1s level of carbon in methane and its fluorine derivatives in the gas phase, and the experimentally determined chemical shift. The chemical shifts are calculated relative to E_B for CH_4, and hence the shift for CH_4 is zero.

Chemical-shift effects are also observed in the solid state, and are often readily explicable in electronic terms, as above. One simple example is the salt sodium azide, NaN_3. The azide ion is linear and can be represented by the structure shown in the margin, in which the two terminal nitrogen atoms are identical, but in a different chemical environment from the central nitrogen. As a consequence, there is a difference in electron densities, as indicated by the signs, which represent partial charges. The N(1s) XPS spectrum of NaN_3 is shown in Figure 18.

$\bar{N}=\overset{+}{N}=\bar{N}$
the azide ion

■ How can the form of the spectrum in Figure 18 be accounted for?

▪ As the spectrum is for the N(1s) level, spin–orbit splitting is ruled out. The two peaks must represent nitrogen in different chemical environments, in accord with the structure of N_3^-. The positively charged, central nitrogen should have the higher binding energy. Moreover, the intensities of the peaks are in the approximate ratio 2 : 1, although in general such simple intensity–abundance relations are not observed in XPS.

The correlation of shift in binding energy with change in electron density suggests that XPS may be put to further use, namely in identifying oxidation states for which the charge density differences may be large. Figure 19 illustrates this effect with the Sb(3d) part of the spectrum of the ionic compound Cs_2SbCl_6. Each of the spin–orbit split peaks is itself composed of two roughly equal components.

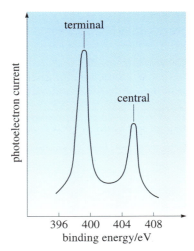

Figure 18 The N(1s) signals in the XPS spectrum of sodium azide.

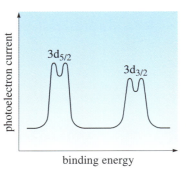

Figure 19 The antimony 3d signals in the XPS spectrum of Cs_2SbCl_6.

■ Given that antimony is in Group V of the Periodic Table, suggest a reason for the presence of two peaks in each of the spin–orbit components in Figure 19.

▪ The formula Cs_2SbCl_6 suggests an average oxidation state of +4 for antimony. This is not consistent with the chemistry of elements in Group V, which are known commonly to adopt the oxidation states +3 and +5. Presumably, the two components arise from antimony(III) and antimony(V) in equal proportions. So Cs_2SbCl_6 is deduced to be a 'mixed-valence' compound.

Unfortunately, not all examples give such conclusive evidence. One oxide of lead, Pb_3O_4, is known from X-ray crystallography to contain octahedrally coordinated lead(IV) atoms and three-coordinate lead(II) atoms within the crystal, and yet the lead XPS spectrum shows no evidence of this. Here, each of the two components of the 4f spectrum (j = 5/2 and 7/2 due to spin–orbit effects) is a single broad peak, possibly the outcome of opposing influences of the inherent difference in oxidation state and environmental (lattice) effects. Thus, XPS sometimes provides a means of detecting different oxidation states, but not always.

Of more interest in catalysis is the chemical shift associated with adsorption. One example is the effect that a small fractional coverage of barium on a zinc surface has on the rate of cleavage of the O=O bond in molecular oxygen. Exposure of the surface of a clean zinc crystal to molecular oxygen gives the O(1s) XPS spectrum in Figure 20a, showing a peak at E_B = 530.2 eV, which is characteristic of ZnO. The effect of barium deposition on the same surface is compared in Figures 20b and 20c. The O(1s) peak shifts to lower binding energy (529.2 eV) due to the formation of BaO, and a further feature appears at 532.0 eV. This is ascribed to the species $O_2^{\delta-}$, which is associated with the formation of a peroxo species (O_2^{2-}) prior to oxygen–oxygen bond cleavage.

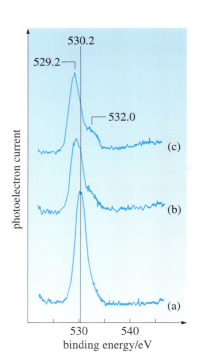

Figure 20 O(1s) XPS spectra, measured with a relatively high resolution, of: (a) a clean zinc crystal surface exposed to oxygen; (b) and (c) doping of the same surface with increasing fractional coverage of barium. Spectra (b) and (c) are displaced vertically for clarity; the base line is a similar PE current in each spectrum.

5.3 Ultraviolet photoelectron spectroscopy (UPS)

Ultraviolet photoelectron spectroscopy, UPS, is usually performed with a helium discharge lamp as the photon source, using the intense emission at 21.2 eV. As the work function, ϕ, for metal surfaces is usually about 4 eV, the kinetic energies of the photoelectrons therefore lie in the range 0–17 eV. Inspection of Figure 3 shows that the path length of electrons from UPS is much greater than that for XPS. For the substrate material, the technique is better described as a bulk rather than a surface method. It provides a useful description of the band structure of materials. For example, Figure 21a includes the UPS spectrum of clean nickel in which the large peaks at lowest binding energies are due to emission from the Ni 3d states in the conduction band. Note that the lowest measured binding energy, at the extreme left of the spectrum, must represent emission of the most weakly bound electrons in the metal, namely those at the Fermi level.

The binding energy of an electron, defined as the energy required to remove the electron to the vacuum level, includes a contribution from the work function (consider, for example, Figure 5). We have seen, however, that ϕ may change on adsorption by up to 1 eV. Uncertainties as to the magnitude of this change are negligible compared with the core level binding energies examined by XPS, but they may be significant in relation to the much smaller values measured by UPS. The problem can be avoided by redefining E_B as the energy difference between a given electron energy level and the *Fermi* level, rather than the vacuum level. In other words, the zero on the E_B scale is now taken as the Fermi level and so the work function is no longer involved. Because the Fermi level can be located at the point in the spectrum where emission is first detected, this provides a more accurate measure of E_B, and UPS spectra are generally presented with E_F as energy zero.

Where UPS comes into its own in surface chemistry is in the study of species that occur at the surface but are absent from the bulk — in other words adsorbates. A single example serves to illustrate the application of UPS and the results it provides in this area. Figure 21 also shows a set of UPS data obtained during a study of the chemisorption of benzene on a nickel surface. In Figure 21a the spectrum of a clean nickel surface is compared with that of the same surface with a chemisorbed layer of benzene. The first thing to notice is the change in the large peak at low binding energies, which confirms what we might well have supposed, namely that the Ni 3d (conduction band) electrons are directly involved in the bonding of benzene to the surface. Of more interest are the peaks that appear at higher energies following adsorption. Focusing on these, the two signals in Figure 21a are subtracted one from another to give the difference spectrum in Figure 21b, which is effectively that of chemisorbed benzene. In order to compare this difference spectrum with the gas-phase UPS spectrum of benzene (Figure 21c), it is necessary to shift the binding energy axis of the benzene spectrum by about 4 eV, due partly to the work function of the metal. However, even when this is done (as it has been in Figure 21c), there is not exact correspondence of the lowest binding energy peak for benzene, the peak labelled π

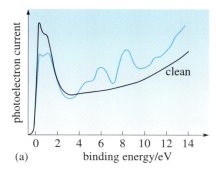
(a)

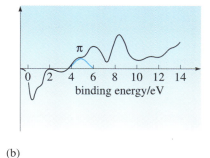
(b)

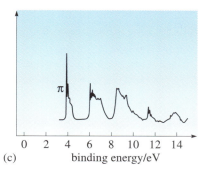
(c)

Figure 21 UPS spectral study of benzene adsorbed on nickel: (a) spectra of the clean nickel surface (black) and the surface with adsorbed benzene (blue); (b) the difference spectrum of the two spectra in (a); (c) the spectrum of gaseous benzene, with the binding energy axis shifted by about 4 eV. (The negative peak in the region 0–2 eV in the difference spectrum (b) results from suppression of the conduction band photoemission when the surface has an adsorbed layer of benzene. The fine structure in the peaks results from vibrational changes, which do not concern us here.) The contribution of the π orbital is shown in blue in part b.

in Figure 21c. This additional shift is caused by the bonding of benzene to the nickel surface, and so identification of the benzene orbital involved reveals something of the nature of this bonding. In benzene, the occupied orbital of highest energy (lowest E_B) is known to be a π orbital in which the electrons are distributed sandwich-like on either side of the carbon plane (Figure 22). If this orbital is involved in bonding to the surface, the implication is that the molecule lies flat on the surface.

We conclude that in favourable cases, such as this one, the mode of bonding of the adsorbate may be deduced from UPS studies.

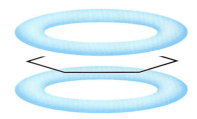

Figure 22 The electron distribution (in blue) in the highest-energy occupied molecular orbital (π) of benzene. The carbon skeleton of the molecule is outlined (black), sandwiched between the two lobes of the molecular orbital.

5.4 Summary of Section 5

1 Photoelectron spectroscopy, PES, is the study of electrons that are emitted when a sample is irradiated with electromagnetic radiation. It provides values of electron binding energies, which can be related to orbital energies.

2 There are two types of PES, namely XPS (using X-ray sources), which can be used to study core and valence electrons, and UPS (using uv sources), which is restricted to the study of valence electrons.

3 XPS provides analytical data and, via the chemical shift effect, information about the chemical environment of the elements at the surface.

4 Interaction between the spin and orbital motion of the electron causes some PES signals to be split into two components.

5 In favourable cases, UPS provides information about the binding of adsorbates on a surface.

SAQ 4 XPS can be used, without the need for interpretation, as a fingerprinting method to identify a good catalyst. Figure 23 shows the rhodium 3d signals from two catalysts consisting of rhodium dispersed on a carbon support; one catalyst is of high activity and the other is inactive. Suggest reasons for the presence of four Rh 3d peaks in these spectra. [*Hint*   A comparison with the antimony spectra in Figure 19 will help.]

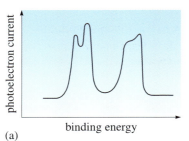

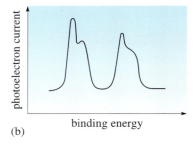

Figure 23 The rhodium 3d signals in the XPS spectrum of a rhodium–charcoal catalyst: (a) a high-activity sample; (b) an inactive sample.

SAQ 5 Steel contains many elements other than iron. Figure 24 shows the XPS spectrum of a steel surface, produced using aluminium K_α X-radiation (1 487 eV). This spectrum plots the *kinetic energies* of the electrons. Some of the peaks in the spectrum occur at 415, 615, 955, 1 203, 1 258, 1 322, 1 375 and 1 424 eV. Use these peak positions and the binding energies in Table 1 to identify the elements present other than iron.

In answering this type of question it is helpful to use the following guidelines:

(i) first try to assign the peaks of lowest E_B;

(ii) for each suspected element, check the spectrum for the presence of confirmatory peaks;

(iii) consider whether each assigned element is a likely chemical constituent of the sample.

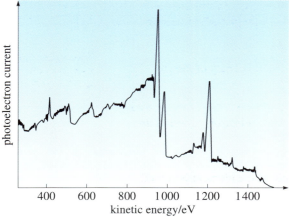

Figure 24 The XPS spectrum of a sample of steel.

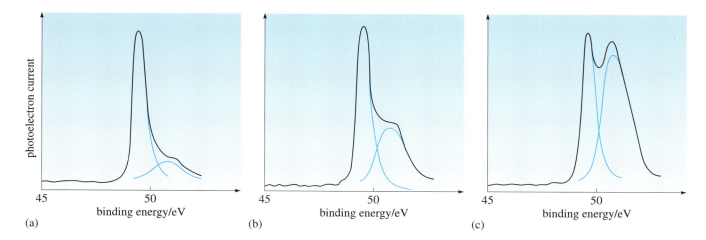

(a) (b) (c)

Figure 25 The change in the Mg(2p) XPS spectrum as a clean magnesium surface is exposed to increasing coverages of oxygen, (a)–(c); spectrum (a) corresponds to the clean surface. In each case, the peaks that make up the composite spectrum are shown in blue.

SAQ 6 Figure 25 shows how the Mg(2p) part of the XPS spectrum of a sample of clean magnesium changes on exposure to small quantities of oxygen. Explain what is happening.

6 AUGER ELECTRON SPECTROSCOPY (AES)

Auger electron spectroscopy is a technique that takes its name from the French chemist Pierre Auger, who first observed the effect in 1923. For experimental reasons, it has developed as a form of spectroscopy only since the 1960s. The Auger effect is most easily understood by considering the fate of the ion formed in the XPS process, as shown in Figure 26.

When photoionization occurs (arrow 1 in Figure 26), the ion is left with a core hole — an electron missing from one of the core levels. The singly charged ion is electronically excited, an inherently unstable situation, since an electron from some higher energy level may relax (fall) into the core hole (arrow 2 in Figure 26). In so doing, a quantum of energy is simultaneously released. This energy may be radiated as a photon or it may be transferred to another electron in the ion. If the energy this electron has gained is greater than its binding energy, it is ejected from the ion (arrow 3 in Figure 26). The overall process is known as the **Auger effect**, and the electron released is an **Auger electron**.

■ How is the energy of the Auger electron related to the frequency of the ionizing photon, as depicted in Figure 26?

□ The two are not related. Following photoionization, the ion is left in an excited state. The subsequent processes of relaxation and Auger electron emission *are inherent to the ion itself*.

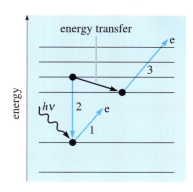

Figure 26 Representation of the various events in Auger electron spectroscopy (not to scale).

Auger electrons thus have energies that are characteristic of the element involved, and so, like XPS, Auger electron spectroscopy provides a means of elemental analysis.

6.1 Experimental aspects

As indicated in Figure 26, Auger electrons are produced when a sample is irradiated with X-rays. They are therefore observed routinely in XPS spectra, and can provide complementary information. However, Auger electrons will be produced by any process that creates a core hole, as in arrow 1 of Figure 26. This can be done more conveniently, and with advantage, by an incident beam of electrons of sufficiently high energy. These incident electrons are called the **primary electrons**, and those observed as a result of scattering or emission processes are called **secondary electrons**. Analysis of the kinetic energies of the secondary electrons can be carried out in a similar way to that in XPS (Figure 10) or by using the experimental arrangement involved in low-energy electron diffraction (Section 8.1).

The spectrum produced in an Auger experiment using a primary electron beam is dominated by electron scattering, which gives a very high background signal. In order to enhance the spectral components due to the Auger electrons, it is common to plot the first derivative of the signal, dI/dE_k, where I is the current of secondary electrons collected from the sample and E_k is the electron kinetic energy (Figure 27). The effect of this is to convert a 'single' Auger peak, which is generally small, into a sharper derivative peak, which is more easily distinguished from the background.

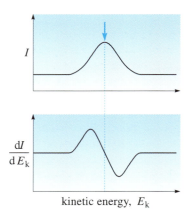

Figure 27 The emission signal (top) in the Auger experiment, and its corresponding derivative peak (bottom). The peak centre is indicated by an arrow.

For most purposes, the primary electron beam is generated with an electron 'gun' that yields energies of about 1 keV, although higher-energy sources (up to 5 keV) can be used. So, as with XPS, the short path length of electrons of these energies makes the technique well suited to surface studies. Again, as with XPS, AES can be used for the identification of any element beyond hydrogen and helium.

Two major differences between XPS and AES arise from the use of the different sources. The primary electron beam may be made very intense, so that AES has a more rapid analysis time and a greater sensitivity; detection of about 10^{13} atom cm^{-2} is possible — less than one-hundredth of a monolayer. However, the advantages of this intensity are sometimes offset by the damaging effect of an electron beam of high energy on the surface under investigation.

A major strength of AES stems from the facility with which a beam of electrons may be focused and moved. At present, AES can be used to examine a spot on a surface as small as 100 nm across and to scan across a surface to build up a compositional map, a topic we return to in Section 11.

Arising as it does from a three-step process, the Auger effect is dependent on more parameters than XPS. For this reason it becomes more difficult than XPS to apply the technique quantitatively.

6.2 Nomenclature in AES

The notation usually adopted for AES is that commonly used in atomic spectroscopy, where the letters K, L, M... identify the electron shells with principal quantum number $n = 1, 2, 3...$ and so on. Numerical subscripts (1, 2, etc.) are attached to these labels in order to take into account the fact that there are different *electron states* within the shell; these electron states are determined by the possible values of j. The simplest example involves one electron in the valence shell, so we shall look at sodium.

The ground-state electronic configuration of sodium is $1s^2 2s^2 2p^6 3s^1$, which gives the following electron energy levels:

K ($1s_{1/2}$), L_1 ($2s_{1/2}$), L_2 ($2p_{1/2}$), L_3 ($2p_{3/2}$) and M_1 ($3s_{1/2}$)

The K shell is not split because there is no spin–orbit coupling for s electrons (Section 5.2.1). The L shell, however has *three* states: again, the s orbital is not split because there is no coupling; this gives the L_1 level. For the electrons in the p orbitals, $l = 1$, so $j = \frac{3}{2}$ or $\frac{1}{2}$ (arising from $j = l \pm s$), giving two states, L_2 and L_3. The three L states are numbered sequentially, with the orbitals from which they arise given in parentheses afterwards. (For sodium, the L_2 and L_3 states are very close in energy, as Table 1 shows, and are commonly shown as one level, labelled $L_{2,3}$.)

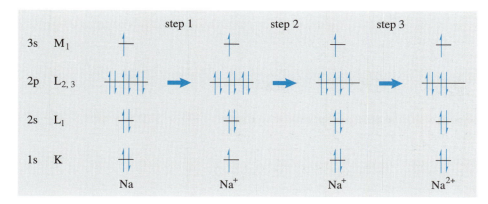

Figure 28 The steps in the emission of an Auger electron from a sodium atom: step 1, the primary electron beam causes a hole to be created; step 2, an $L_{2,3}$ electron falls into the hole, causing energy to be released; step 3, this energy causes the emission of an Auger electron from the $L_{2,3}$ level.

The value of this notation is that it allows us to give a precise description of the processes involved in producing a given Auger electron. Take the process for sodium outlined in Figure 28, for example. In the first step (arrow 1 in Figure 26), the primary beam causes ionization from a core level, K, producing a sodium ion, Na⁺. In the second step, a 2p electron falls from the $L_{2,3}$ level into the hole created by the initial ionization (arrow 2 in Figure 26), and this releases sufficient energy to simultaneously eject a further electron (the Auger electron) from the $L_{2,3}$ level (arrow 3 in Figure 26), leaving a doubly charged Na²⁺ ion. The Auger electron is then denoted by three letters, placed in order according to its history: (a) the hole produced by the initial ionization; (b) the source of the electron filling the hole; (c) the source of the Auger electron.

■ Write down the label for the Auger electron emitted in Figure 28.

▪ The label is $KL_{2,3}L_{2,3}$.

Many Auger transitions involve the valence shell electrons. For an isolated atom, the electrons are identified by energy-level labels as described above; thus in the example in Figure 28 the electron from the valence shell of sodium would be labelled M_1. However, things become rather more complicated in the case of compounds, where the atomic orbitals interact to form bands or molecular orbitals. To get around this complication, electrons from valence orbitals in molecules are simply denoted by the capital letter, V, in this notation.

■ What is the kinetic energy of the Auger electron, E_{Auger}, for the process $KL_1L_{2,3}$ in sodium. (It helps to write out the sequence of events as in Figure 28, noting where energy is absorbed and where it is released.)

▪ The set of transitions corresponding to the $KL_1L_{2,3}$ Auger process is shown in Figure 29. When the L_1 electron relaxes into the hole in the K shell, the energy released is $(E_K - E_{L_1})$. In the ionization of the $L_{2,3}$ electron, an amount of energy $E_{L_{2,3}}$ is used. So the difference between these energies is the kinetic energy with which the Auger electron escapes from the atom:

$$E_{Auger} = E_K - E_{L_1} - E_{L_{2,3}} \qquad (4)*$$

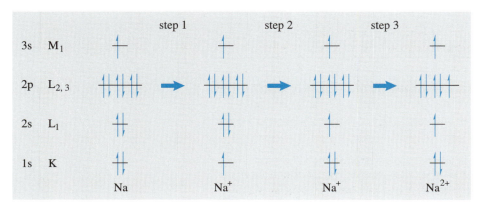

Figure 29 The Auger process represented by the notation $KL_1L_{2,3}$.

* The binding energies E_K, E_{L_1} and $E_{L_{2,3}}$ are ionization energies of the sodium *atom*. In the last step of the Auger process, the electron is lost from a *singly charged ion*, for which the energy levels are different: this introduces an error into equation 4 (which can be corrected for). However, the error is not large enough to prevent the equation being useful for the identification of observed Auger transitions.

In a more general sense, equation 4 may be written as

$$E_{\text{Auger}} = E_X - E_Y - E_Z \qquad (5)$$

where the internal transition occurs between levels Y (upper) and X (lower), and the emission of the Auger electron occurs from level Z.

Notice that equation 5 highlights the important feature of Auger spectroscopy alluded to earlier.

> The energy of the Auger electron is independent of the energy of the ionizing process (*radiation or primary electron*) and depends only on the binding energies of the three levels involved.

Thus in XPS, where Auger electrons occur, the Auger features in the spectrum can be readily identified by changing the energy of the X-ray source. The positions of the Auger peaks will be unaffected, whereas the XPS peaks will be shifted in energy by the energy difference of the two sources.

6.3 Summary of Section 6

1 Auger electron spectroscopy, AES, is the study of electrons emitted by excited ions that are generated either by exposure to electron beams or during photo-ionization.

2 The Auger electron is emitted when an excited ion relaxes. It is labelled according to the energy level from which it derives and the two levels involved in the relaxation.

3 Energies of Auger electrons are independent of the energy of the ionizing process.

4 AES is essentially an analytical method of high sensitivity.

SAQ 7 Summarize the relative advantages and disadvantages of AES *vis-à-vis* XPS by placing ticks or crosses where appropriate in the following table. Place a 'd' where you think there is some difficulty in using the technique.

	XPS	AES
element identification		
sensitivity		
analytical speed		
quantitative analysis		
low surface damage		
spatial information		
chemical environmental effects		
wide range of elements		

SAQ 8 Figure 30 is the Auger spectrum of the surface of a sample of metallic nickel. Use the data in Table 1 to identify the elements that give rise to the Auger signals labelled A and B. Note that these are both light elements in the second Period of the Periodic Table (Li to Ne).

> STUDY COMMENT   If you have not already watched video band 6 (*Electron spectroscopy*), this is an appropriate point to do so. The sequence examines the application of XPS and AES to a particularly important catalytic reaction—the Haber–Bosch process for the synthesis of ammonia.

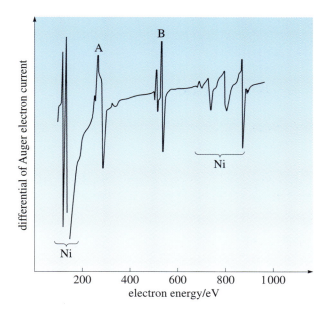

Figure 30 An Auger spectrum of a sample of nickel.

7 THE STRUCTURE OF SURFACES

As we investigate the chemistry of surfaces more closely, the need for a more detailed description of surface structure increases. Many studies are motivated by the quest to understand catalysis. Although commercial catalysts are often supported or polycrystalline materials, commonly in the form of compacted powders, the scientific understanding of catalysis depends crucially on the knowledge of the behaviour of the surfaces of single crystals.

We begin with the simplifying assumption that the surface of a crystal is an exposed plane of atoms whose structure is identical with that of the corresponding plane within the crystal. However, this assumption is rarely valid, as you will see in Topic Study 2, Part 2.

STUDY COMMENT   In this Section we examine the structures of two simple types of solid, which were introduced in the Second Level Inorganic Course. We introduce a notation used to describe the arrangement of atoms on the surface of crystals of these solids. For each of the two structures we concentrate on just three kinds of surface, for which the arrangement of atoms is depicted in the fold-out summary sheet. The key points in this Section are summarized in Section 7.4. Note that you will not be expected to derive the structures of planes in three-dimensional crystals for these or any other types of solid.

7.1 Crystal structures of metals: crystal planes and crystal surfaces

With few exceptions, metals have one of the three structures known as **body-centred cubic** (*bcc*), **face-centred cubic** (*fcc*) — which is also called **cubic close packed** (*ccp*) — and **hexagonal close packed** (*hcp*). Three-dimensional unit cells of two of these structures are shown in Figure 31. The diagrams at the bottom of the Figure are described as space-filling because the balls representing atoms fill the space as far as possible, with nearest neighbours touching. Equivalent representations, shown at the top in Figure 31, are called lattice point models. Here each atom is represented by

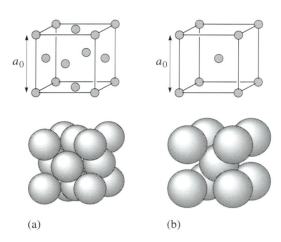

Figure 31 Two of the structures commonly adopted by metals: (a) face-centred cubic; (b) body-centred cubic.

a small circle, and the structure of the lattice is denoted by the thin lines. As indicated in Figure 31, the size of the three-dimensional unit cell in these cubic structures is characterized by the unit cell length or **unit cell parameter**, which we represent by the symbol a_0. In this Section we examine how these structures, which have been deduced by X-ray crystallography, may be visualized as sets of planes or layers containing metal atoms arranged in regular arrays. We concentrate on those planes in which the density of atoms is high, because these have relatively simple structures and tend to form the surfaces of crystals.

In focusing on a particular set of planes in a crystal, we need a means of describing how this set, or any one of its planes, lies in relationship to the three-dimensional unit cell. The first step is to define three axes. In the case of cubic structures, these run along the edges of the unit cell and are drawn according to the following convention, as in Figure 32a.

> Draw the z axis on the page vertically upwards. The positive directions of the x and y axes are then as depicted in Figure 32a; that is, the x axis is perpendicular outwards from the page, and the y axis is horizontal on the page, to the right. However, it should be recognized that in order to draw diagrams with good perspective, it is often necessary to rotate this axis system about the z axis (as in Figure 32b and c).

The set of planes can now be identified by three integers, h, k and l, which are related to the distances from the origin at which any one plane in the set cuts the axes x, y and z, respectively. The integers are enclosed in brackets, (hkl), and are known as the **Miller indices** of the plane or planes in question.

We shall not go into details of the procedure for working out Miller indices, except to note two points:

(i) if a plane is parallel to an axis, the corresponding h, k or l value is zero, so any one or all of the planes in Figure 32b would have Miller indices of the type $(hk0)$ because they all lie parallel to the z axis;

(ii) if the intersection of a plane with an axis has a negative value, this is indicated by placing a 'bar' over the corresponding h, k or l value, so the plane in Figure 32c would have Miller indices of the type $(h\bar{k}l)$, because it cuts the y axis at a negative value.

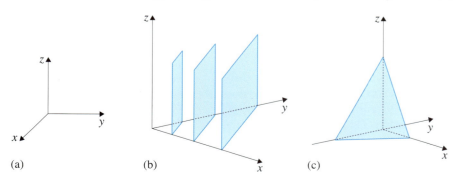

Figure 32 (a) Definition of axes; (b) planes with Miller indices of the type $(hk0)$; (c) planes with Miller indices of the type $(h\bar{k}l)$.

In addition to identifying planes, it is often necessary to specify a particular *direction* running through a crystal lattice; for example, the direction MN in Figure 33. The procedure for doing this is as follows:

(i) through the origin, draw a line PQ parallel to MN;

(ii) choose some convenient point on this line and determine its coordinates, r, s and t (in the x, y and z directions, respectively), using the unit cell length, a_0, as the unit of measurement;

(iii) if necessary, convert the coordinates to the set of smallest integers in the same ratios, u, v and w.

Written in square brackets, [uvw], with any negative values indicated by a 'bar', these integers represent the direction of the line PQ, or any other line parallel to it, such as MN. Figure 33 illustrates a specific example.

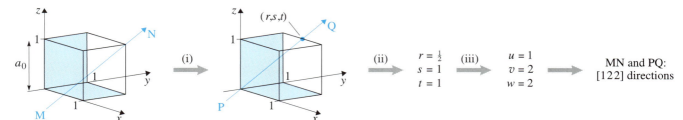

Figure 33 Steps involved in the determination of a direction, MN, in a cubic lattice, of unit cell length a_0.

SAQ 9 Use the procedure described above to determine the directions AB, AC and AD in the cubic crystal shown in Figure 34.

We now examine the three sets of planes in the *fcc* system, which have the highest atom densities. These have the Miller indices (100)*, (110) and (111), and their relationships to the cubic unit cell are shown in Figure 35.

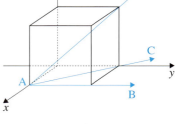

Figure 34 Three directions, labelled AB, AC and AD in a cubic crystal.

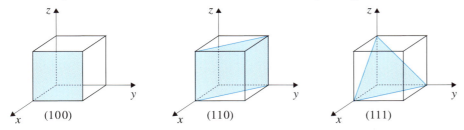

Figure 35 Planes in a cubic crystal with Miller indices (100), (110) and (111), and their relationships to the cubic unit cell.

(a) *fcc* (100) planes

The (100) is the most obvious set of planes in a cubic structure, and includes the front and rear faces of the cube, represented by open and grey circles in Figure 36. It is most important to realize, however, that halfway between these two planes there is a further plane (blue circles in Figure 36), which has an identical atomic arrangement. It has been shifted, of course, by $\frac{1}{2}a_0$ in the x direction, but it is still a (100) plane. Thus, the interlayer spacing between each plane in the complete set is $\frac{1}{2}a_0$.

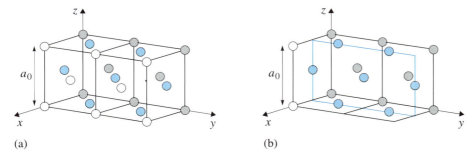

Figure 36 (a) Two unit cells of the *fcc* structure; (b) the same structure, but with the front faces of the cells removed to emphasize the atomic arrangement on the intermediate (100) plane (blue circles).

* Spoken as 'one, nought, nought', etc.

It is obvious from Figure 36 that there are two other sets of planes similar to (100), namely those parallel to the sides of the cubic cell and those parallel to its top and base. (The corresponding Miller indices are (010) and (001), respectively.) From the point of view of surface science, all three sets are equivalent, and it is useful to depict them (or indeed, any collection of equivalent planes) by *curly* brackets round the indices. So, the notation {100} indicates any plane with the same atomic arrangement as (100).

We now examine the structure of the surface of a *fcc* crystal that has been cut to expose a (100) plane. The *two*-dimensional lattice of the top atomic layer is shown in Figure 37, on which one face of the three-dimensional unit cell is outlined.

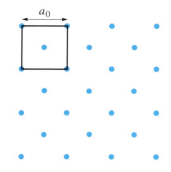

Figure 37 The surface of a *fcc* crystal cut to expose a (100) plane.

■ Is this unit cell the smallest repeating unit from which the two-dimensional lattice in Figure 37 can be constructed?

■ If the pattern in Figure 37 is rotated by 45°, it appears as a simple square array. The smallest repeat unit, the two-dimensional **surface unit mesh**, is clearly the smaller square outlined in Figure 38.

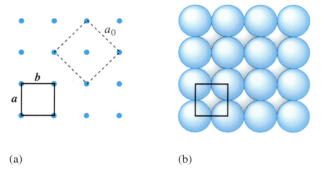

(a)   (b)

Figure 38 The (100) surface of the *fcc* structure in both (a) the lattice point and (b) the space-filling representations. The face of one three-dimensional unit cell is indicated by dashed lines in part (a). *a* and *b* are the two-dimensional unit mesh vectors (defined in the text), with magnitudes $a = b = \sqrt{2}a_0/2$.

Notice that the length of one side of this small square is a half of the diagonal of the square derived from the three-dimensional unit cell (also shown in Figure 38a), that is, $a = b = \sqrt{2}a_0/2 = a_0/\sqrt{2}$.

As you might expect, for drawing surface unit meshes such as that in Figure 38, there is a convention to be followed, which is detailed in Box 1.

Box 1 Convention for drawing a surface unit mesh

1 The sides of the mesh are represented in magnitude and *direction* by two **unit mesh vectors**, *a* and *b*, chosen so that $a \leqslant b$ (where *a* stands for the magnitude, or length, of ***a***). Note that vectors are always represented by symbols printed as bold italics.

2 The ***b*** vector is always drawn horizontally, from left to right.

3 The ***a*** vector is drawn downwards, either vertically or, if necessary, to the left (the angle between ***a*** and ***b*** is then always $\geqslant 90°$).

(b) *fcc*(110) planes

A lattice point representation of two unit cells of the *fcc* structure is shown in Figure 39a, with the 110 planes shaded in blue. As in the (100) case above, it is important to realize that halfway between each of these planes is another plane, identical apart from being shifted relative to the coloured set, by $\tfrac{1}{2}a_0$ in both the *x* and *y* directions (Figure 39b). Thus, the interlayer spacing in the complete (110) set is $\sqrt{2}a_0/4$.

The surface of a crystal cut along a (110) plane is represented by the two-dimensional lattice in Figure 40a. Here, the simplest repeating unit (that is, the unit mesh) is rectangular. Note that to conform with the convention discussed above, the general structure depicted by Figure 39 has been rotated through 90° about the *y* axis in order to draw Figure 40a so that the longer ***b*** side of the mesh lies horizontally to the right. The perspective view of the space-filling model (Figure 40c) emphasizes that the surface consists of parallel 'channels' separated by close-packed atomic rows.

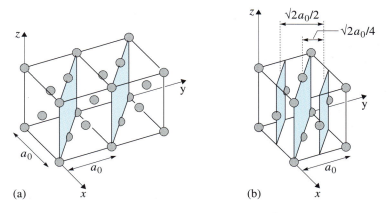

Figure 39 (a) Two unit cells of the *fcc* structure with the (110) planes shaded in blue; (b) a single unit cell with the intermediate (110) planes also emphasized in blue. Notice that the interlayer spacing in the complete (110) set is $\sqrt{2}a_0/4$.

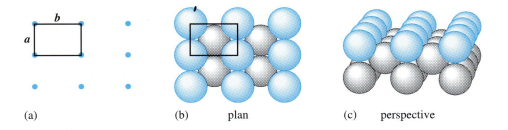

Figure 40 The (110) surface of the *fcc* structure in (a) the lattice point representation and the space-filling representation (showing both (b) plan and (c) perspective views). The lengths of the sides of the two-dimensional unit mesh are $a = \sqrt{2}a_0/2$ and $b = a_0$. Notice that in the space-filling representation the surface layer of atoms is depicted in blue and that deeper layers are shaded grey. This is a convention we adopt in all subsequent figures that show surfaces of three-dimensional structures.

(c) *fcc*(111) planes

The third type of *fcc* plane that we consider is shown (as blue atoms) in relation to the three-dimensional unit cell in a space-filling model in Figure 41a, together with the equivalent lattice-point representation in Figure 41b.

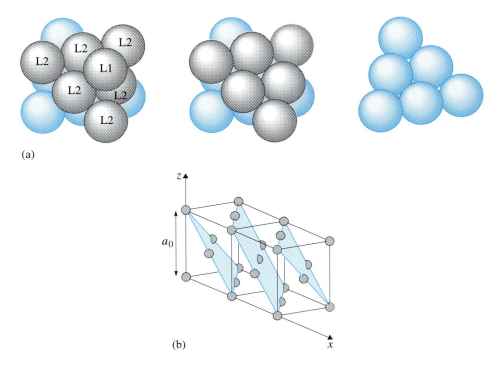

Figure 41 (a) The *fcc* structure, showing one unit cell with successive layers removed to show a (blue) (111) layer; (b) lattice-point representation of two *fcc* unit cells, showing the location of the (111) planes (blue shading).

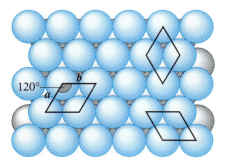

Figure 42 The first and second atomic layers of the $fcc(111)$ structure. Three possible unit meshes are shown, with the conventional one drawn on the left. The mesh sides are $a = b = \sqrt{2}a_0/2$.

The structure of the (111) surface is most easily envisaged using the space-filling picture, as in the extended view in Figure 42. Notice that the atomic array has hexagonal symmetry. In addition, each atom touches all its immediate neighbours so that the available space is occupied as far as possible. It is for this reason that the three-dimensional *fcc* structure is known also as *cubic close packed*.

The surface unit mesh in this case is a parallelogram with equal sides, $a = b = \sqrt{2}a_0/2$, and angles of 120° and 60° (as required by the hexagonal symmetry). Three alternative ways in which the mesh might be drawn are included in Figure 42, but only that shown on the left conforms with convention. (This is the first example we have encountered where the *a* vector must slope down to the left.)

The three surfaces described so far are chosen because they have high thermodynamic stability, making them the ones most likely to constitute the surfaces of the small metal crystals (crystallites) found in many commercial catalysts. They

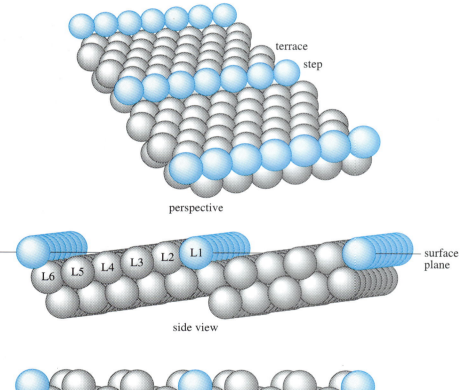

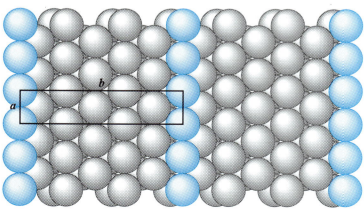

Figure 43 Perspective, side and plan views of a $fcc(332)$ surface, showing the appearance of terraces and steps. In each representation the atoms in the surface layer are shown blue. Successive atomic layers, from the surface plane downwards, are labelled on the side view, and the surface unit mesh is indicated on the plan view.

have high atom densities and low Miller indices, often being referred to as 'the **low-index planes**'. By contrast, a surface formed of a plane with high indices has a *low* atom density in its outermost layer, as the *fcc*(332) example in Figure 43 illustrates. Surfaces of low atom density, such as this, are called **open** or **rough**. All 'high-index' surfaces can be viewed as a series of *terraces* of low-index structure, separated by *steps* one atom in height and also of low-index type.

■ What types of low-index structure do the terraces and steps in Figure 43 appear to have? (Notice that a step consists of a row of blue atoms and the row immediately below it; this is best seen in the perspective view.)

▪ Comparison with the three low-index planes considered above shows that both the terraces and the steps are of the {111} type.

Although metal crystallites may have low-index *faces*, it is often suggested that the sites of highest catalytic activity are located at atomically rough features on the crystallite surface, such as the edges along which two faces meet. The reactivity of high-index surfaces is then of considerable interest because steps provide models for features of this type.

Figure 43 illustrates an important point concerning many surfaces, particularly those of high indices. By no means all of the atoms that will be significant for adsorption and catalysis lie within what, in geometrical terms, would be defined as the surface plane — that is, the first atomic layer. In the *fcc*(332) case the first six layers (L1–L6) are directly exposed to the gas phase, and even with *fcc*(110) the same is true for atoms in the second layer (Figure 40).

Before closing this Section, it will be useful to illustrate the three low-index planes for the *bcc* structure; these are shown in Figure 44. There are close similarities between the *bcc*(100) and *fcc*(100) surfaces, and between *bcc*(110) and *fcc*(111), but the surface equivalent to the 'channelled' *fcc*(110) is *bcc*(211), which is also shown in Figure 44. Notice that the *bcc*(111) surface has a relatively open atomic array, which has hexagonal symmetry.

Figure 44 Structures of three low-index planes of the *bcc* structure represented as (a) lattice points and (b) space-filling structures. For each of the (100), (110) and (111) surfaces, the two mesh sides are of equal length, with the following values: (100) a_0; (110) $\sqrt{3}a_0/2$; (111) $\sqrt{2}a_0$. The (211) surface is also shown because it is equivalent to the very similar 'channelled' *fcc*(110) surface. The mesh sides are of length $a = \sqrt{3}a_0/2$ and $b = \sqrt{2}a_0$. Perspective views of the (111) and (211) surfaces are also shown; in the perspective view of the (111) surface, the layers are numbered (L1–L4) according to depth.

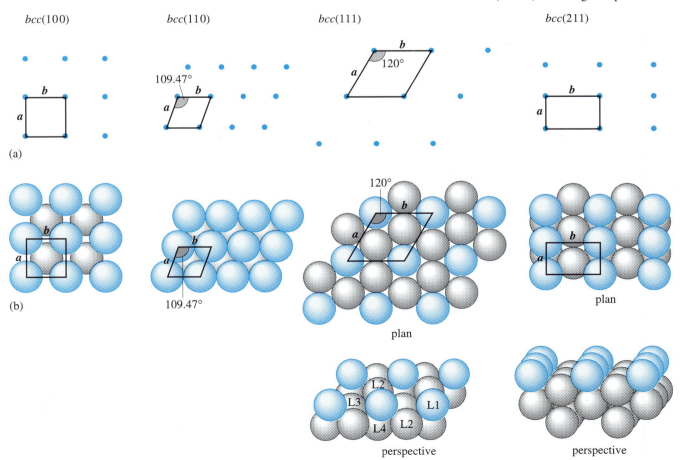

- Which of the three low-index types of *bcc* surface plane, (100), (110) or (111), appears to have the highest density of atoms?

- It is in fact the (110) plane, where in the space-filling model the atoms are seen to be almost close packed.

This illustrates again that it is the low-index planes that have high atom densities, although the *fcc* and *bcc* planes of *lowest* indices, namely (100), are not the *most* dense. Surface density is estimated as follows.

The area of a parallelogram involves the lengths of the sides and the angle between them, and so the area of the surface unit mesh of an (*hkl*) plane is

$$S_{hkl} = ab \sin \psi \qquad (6)$$

where ψ is the mesh angle between **a** and **b**. The atom at each corner of one mesh is shared with three other meshes. Each mesh contains contributions from four atoms, so the total number of surface-layer atoms in a single mesh is one. Thus, the metal atom density in the top layer at the surface is

$$\textbf{surface density} = 1/S_{hkl} \qquad (7)$$

which becomes $1/ab$ in the case of a square or rectangular mesh, where $\psi = 90°$, and $2/\sqrt{3}ab$ for a hexagonal mesh, where $\psi = 120°$.

> **STUDY COMMENT** In this subsection we have focused on three low-index planes of the *fcc* and *bcc* structures. For convenient reference, the fold-out sheet at the end of the Block shows the structures of these planes with their unit meshes, and the sizes of these meshes expressed in terms of the three-dimensional unit cell parameter, a_0.

SAQ 10 The three-dimensional unit cell parameter of the *bcc* metal tantalum is $a_{0,\text{Ta}} = 0.330$ nm. By referring to Figure 44, calculate the surface atom density in the Ta(110) plane.

SAQ 11 Sketch on Figure 45 the unit mesh for each of the three types of surface shown. (It is best to view the models as geometrical 'wallpaper' patterns and look for the simplest repeating unit, which, of course, should conform if possible with the convention for depicting unit meshes.)

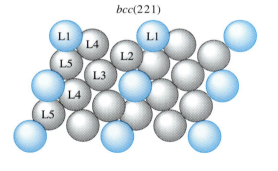

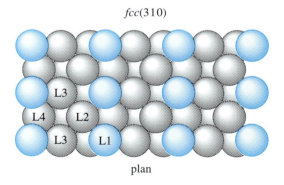

plan

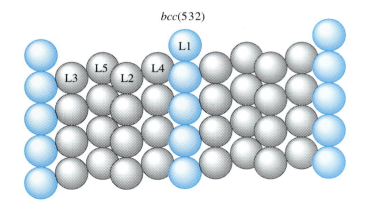

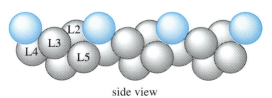

side view

Figure 45 Space-filling plan views of the *bcc*(221) and *bcc*(532) surfaces, and a plan and side view of the *fcc*(310) surface.

7.2 Notation for the structure of adsorbed layers on surfaces

We now turn to the description and notation of surfaces that include adsorbate layers. Our discussion of three-dimensional structures implies that the regularity of a three-dimensional crystal, called the **substrate structure**, continues to its surface. We have seen that whereas a solid may be represented by a three-dimensional lattice, with a three-dimensional unit cell, a *surface plane* is represented by a two-dimensional lattice with, of course, a two-dimensional repeating unit — the unit mesh. This Section introduces a notation for adsorbed layers, which are assumed to be complete and perfect arrays, an oversimple but necessary constraint. We shall also assume for the moment that the adsorbate-covered surface is purely two dimensional, although, as with clean surfaces, atoms or molecules lying beneath the top layer may well also be important.

To describe the unit mesh of the adsorbate, the mesh of the clean surface (the substrate) is used as a reference, with the lengths of its sides, a and b, each taken as one unit of measurement. In many cases the sides of the **adsorbate mesh** lie in the same directions as those of the substrate, and have lengths that are integral multiples of a and b — that is, ma and nb (where m and n are integers). The adsorbed layer is then said to have an $(m \times n)$ unit mesh (pronounced 'em by en') and the structure is fully characterized by specifying also the metal, the surface plane, and the nature of the adsorbate, for example: Cu(110)(5 × 2)–S, where the adsorbate is sulphur. Using this notation, it follows that the mesh of the clean substrate is *always* (1 × 1), even when its sides a and b are *not* of equal length.

We shall first consider two simple examples. Oxygen adsorbs dissociatively on the (100) surface of palladium (an *fcc* metal) to give the structure shown in Figure 46a. The sides of the adsorbate mesh are twice as long as those of the substrate mesh, and so we have a (2 × 2) layer or, in full, Pd(100)(2 × 2)–O. On the (111) plane of platinum (which is another *fcc* metal), hydrogen adsorbs dissociatively to give a mesh identical to that of the substrate, and here the structure is Pt(111)(1 × 1)–H (Figure 46b).

In both of the above cases, the adsorbate mesh sides were equal, but suppose this was not so. We might consider, for example, a (1 × 2) layer. The following question then arises: is this the same as a (2 × 1) layer? Provided the *substrate mesh sides are of equal length* ($a = b$), the answer is yes. Figure 47a shows that, despite the rotation of 90°, the two adsorbate meshes are identical on an *fcc*(100) surface. The structural relationship between the adsorbed species and the metal surface atoms is the same in both cases.

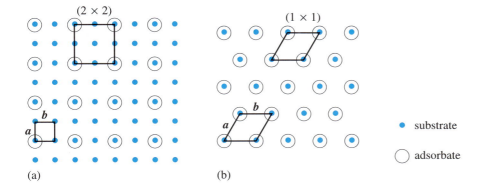

Figure 46 The surface structures of: (a) oxygen adsorbed on Pd(100); (b) hydrogen adsorbed on Pt(111).

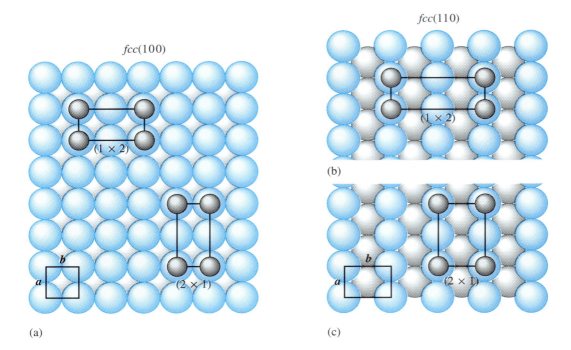

(a)

(b)

(c)

Figure 47 (1×2) and (2×1) layers as examples of adsorbate meshes with unequal sides. (a) When the substrate (1×1) mesh has *equal* sides, for example $fcc(100)$, the two adsorbate structures differ only by a rotation, in this case of 90°. (b) and (c) If, however, the substrate mesh sides are *unequal*, as with an $fcc(110)$ surface, the two adsorbed layers have quite different structures.

When, however, the substrate mesh sides are *not* of equal length ($a < b$), the structures of (1×2) and (2×1) layers will be distinctly different. The simplest example is that of the $fcc(110)$ surface, as shown in Figure 47 (b and c). In the (1×2) structure (Figure 47b) the adsorbed species are close packed in the a direction, along the 'rows' of surface metal atoms, and double spaced in the b direction, across the 'channels'. For the (2×1) layer, the situation is reversed (Figure 47c).

Figure 47c illustrates a further important point. The convention used when representing clean-surface (substrate) meshes, that the *longer side*, b, should be drawn horizontally, does not have to be followed in the adsorbate case. An experimental structural determination will tell us in which directions the longer and shorter adsorbate mesh sides lie, and it is this information alone that dictates how we must draw the mesh on a diagram of the clean surface.

In the cases discussed so far, the adsorbate mesh has been aligned with that of the substrate — the sides of both meshes have run along the same directions — but it is also possible for the meshes to be rotated with respect to one another. One common example is found with adsorption on the (100) planes of fcc and bcc metals, such as dissociative nitrogen adsorption on Fe(100) (Figure 48).

- How does the size of the 'unit mesh' outlined in Figure 48 compare with that of the Fe(100) clean surface?

- It is twice as large in both the a and b directions, and so we might be tempted to label it (2×2).

The mesh, however, contains a nitrogen atom at its centre and so a more appropriate label would be $c(2 \times 2)$, the 'c' standing for 'centred'. The structure is, in fact, frequently known as 'centred two-by-two', but strictly this is incorrect. As seen in Figure 49, the *simplest* repeating unit, and hence the true mesh, is the dashed square on the right.

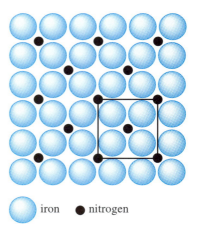

Figure 48 The sites of adsorbed nitrogen on the Fe(100) surface. Iron is a bcc metal. A space-filling representation of the $bcc(100)$ surface is shown in Figure 44b.

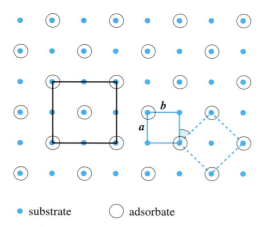

Figure 49 The surface structure of nitrogen adsorbed on Fe(100), showing the centred adsorbate unit mesh of Figure 48, and the true adsorbate unit mesh (dashed blue lines). The small blue mesh is for the substrate Fe(100) plane.

■ What are the dimensions of this smaller adsorbate mesh, and through how many degrees is it rotated with respect to the substrate unit mesh (the angle marked in Figure 49)?

▫ The side of the mesh is a *diagonal* of the underlying substrate mesh, and so has sides of length $\sqrt{2}a_0$. The angle of rotation is 45°.

The correct (though not universally used) notation for this structure is thus Fe(100)($\sqrt{2} \times \sqrt{2}$)R45°–N, where R45° denotes the rotation of the adsorbate mesh relative to the substrate mesh. This shows clearly its relationship in both *size* and *orientation* to the substrate.

SAQ 12 Figure 50 shows some simple substrate structures (as dots) with adsorbed species, shown as circles. In each case the substrate unit mesh is outlined in black at the top left of the figure. Give the notation that describes each of the adsorbed layers.

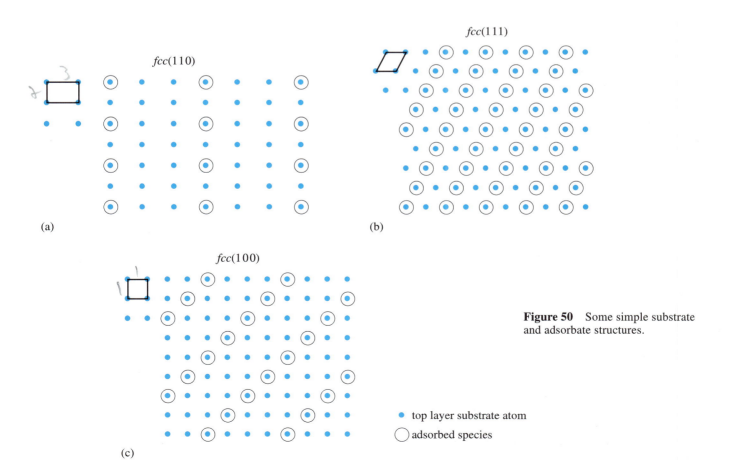

Figure 50 Some simple substrate and adsorbate structures.

7.3 Fractional surface coverage

In Block 5, the fractional surface coverage, θ, of an adsorbate was defined as

$$\theta = \frac{\text{amount adsorbed}}{\text{monolayer capacity}} \tag{8}$$

where the monolayer capacity is the amount contained in a single complete layer, either physically or chemically adsorbed under the same conditions of temperature and pressure. Given this definition, θ can never be greater than unity. It was also mentioned that a more precise definition of θ becomes possible in the particular case of single crystal surfaces. For a single adsorbate, it can be taken as the ratio of the number of species adsorbed to the number of substrate atoms in the *top* atomic layer of the solid (both quantities relating to unit surface area).

If adsorbed species occur only at the corners of the unit mesh, each is shared between four meshes. In other words, there is a total of one adsorbed species in a single mesh. If the mesh is $(m \times n)$, with dimensions ma and nb, its area will be mn times *greater* than that of the (1×1) clean substrate mesh, which likewise contains just one top-layer substrate atom. Therefore, within unit area of the surface the number of adsorbed species will be *smaller* than the number of top-layer metal atoms by a factor of $(1/mn)$. The **fractional surface coverage** is then simply

$$\theta = \frac{1}{mn} \tag{9}$$

It is important to note that it is quite possible for the number of adsorbate species within the unit mesh, call it x, to be greater than one. In that case, the definition of fractional coverage is modified to

$$\theta = \frac{x}{mn} \tag{10}$$

which can lead to a value of θ *greater* than one (see SAQ 13).

As discussed previously, the atomic density in the top substrate layer is $1/S_{hkl}$, where S_{hkl} is the clean-surface mesh area. So, if required, the density of the adsorbate can be calculated as

$$\text{density of adsorbate} = \theta \times \frac{1}{S_{hkl}} \tag{11}$$

Up to this point we have mostly shown adsorbed species as being located directly on top of substrate atoms, in 'terminal' adsorption sites (for example, see Figure 51a) but, of course, other types of site are also present on any surface. In Figures 51b and 51c the adsorbate mesh is (2×2), as in Figure 51a, but the sites occupied are 'twofold bridge' and 'threefold hollow', respectively. All three structures would be labelled (2×2), the difference between them being in the positions of the metal surface atoms relative to the adsorbed species. This emphasizes again the importance of the three-dimensional surface unit cell, involving one or more of the uppermost substrate layers, as opposed to the purely two-dimensional unit mesh. As we shall see in the next Section, the technique of low-energy electron diffraction provides a means of locating the atoms within the surface unit cell, and thus of distinguishing one structure from other possible structures such as those in Figure 51.

SAQ 13 Hydrogen adsorbs dissociatively on the (110) surface of the *fcc* metal rhodium to give a series of different structures with increasing coverage, two of which are shown in Figure 52.

In each case identify the surface unit mesh (examine the structure in purely geometrical terms, looking for the smallest unit that, when repeated, will generate the *complete* pattern). Label the mesh using the $(m \times n)$ notation, and determine the number of adsorbed species contained within it. Hence, calculate the fractional surface coverages, θ. Given that the three-dimensional unit cell parameter is $a_{0,\text{Rh}} = 0.380$ nm, also calculate the hydrogen atom surface density at maximum coverage.

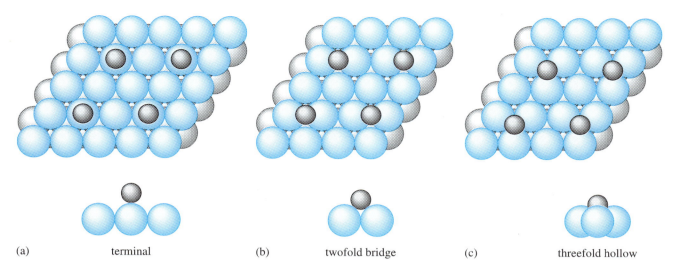

(a) terminal (b) twofold bridge (c) threefold hollow

Figure 51 Three possible types of adsorption site on the *fcc*(111) surface. In each case the adsorbate is arranged in a (2 × 2) unit mesh.

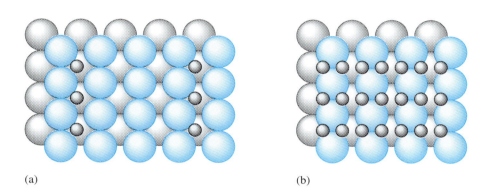

(a) (b)

Figure 52 Structures formed by hydrogen atoms (small black circles) adsorbed on Rh(110) (a) at low coverage and (b) at maximum coverage.

7.4 Summary of Section 7

1 Within crystals the atoms are arranged in regular arrays.

2 An array may be represented by a three-dimensional lattice of points or by a space-filling model.

3 The smallest unit from which a three-dimensional array may be constructed is called the unit cell. For cubic systems, such as *fcc* and *bcc*, the unit cell length (a_0) is the same in the three directions, x, y and z.

4 Within the crystal the atoms can be envisaged as lying on planes.

5 Crystal planes are defined by Miller indices, (hkl). Sets of equivalent planes (that is, with the same atomic arrangement) are denoted by $\{hkl\}$.

6 A direction running through a crystal lattice is denoted by [uvw], according to the procedure illustrated in Figure 33 (Section 7.1).

7 Surfaces of crystals may be represented by a two-dimensional array of points or by a space-filling model.

8 The smallest unit from which a two-dimensional array may be constructed is called the unit mesh.

9 The sides of the unit mesh are represented by vectors ***a*** and ***b***, of length a and b, respectively, and are always chosen so that $a \leqslant b$. When drawing the unit mesh, the ***b*** vector is shown pointing horizontally to the right, with the ***a*** vector pointing downwards and, if necessary, leftwards. For reference, the structures and unit meshes of some low-index planes in the *fcc* and *bcc* systems are illustrated on the fold-out sheet at the end of the Block, together with details of the mesh parameters a and b.

10   Many properties of a surface depend on atoms below the top layer (the geometrical surface plane), and so it is often necessary to consider a three-dimensional 'surface unit cell', extending some distance beneath the unit mesh.

11   Crystals tend to form with surfaces that have relatively high atom densities.

12   Surfaces of high Miller indices can be regarded as a series of steps and terraces, which have the structure of low-index planes.

13   Adsorbates usually adopt regular arrays on a surface, determined by their interaction with the substrate.

14   An adsorbed layer is represented by a unit mesh and described by integers m and n, multiples of the lengths (a and b, respectively) of the substrate mesh. The adsorbate mesh may be rotated relative to the substrate mesh. In general, the notation used to describe the adsorbate structure can be summarized as shown in Box 2.

Box 2   Notation used to describe adsorbate structures

$$M(hkl)(m \times n)RX°\text{–}A$$

where hkl is the surface plane of the metal (M) substrate,

$m \times n$ is the size of the adsorbate unit mesh with respect to the substrate unit mesh,

$X°$ is the orientation of the adsorbate unit mesh with respect to the substrate unit mesh, and

A is the identity of the adsorbate.

---

15   In general, the fractional surface coverage, θ, is defined as the amount adsorbed divided by the monolayer capacity. For single crystal surfaces, θ can be defined more precisely as the ratio of the number of adsorbed species to the number of substrate atoms in unit surface area of the top layer of the solid.

16   For an adsorbate on a single crystal surface (hkl), the surface density of the adsorbate can be calculated as θ/S_{hkl}, where S_{hkl} is the area of the substrate unit mesh: $S_{hkl} = ab \sin \psi$ (ψ being the mesh angle between a and b).

8 LOW-ENERGY ELECTRON DIFFRACTION (LEED)

Many of the properties of electrons are better described by modelling them as waves rather than as particles. In particular, a beam of electrons is diffracted by the regular array of atoms in a metallic crystal according to an associated wavelength determined by the kinetic energy of the electron. The wave nature of electrons is exploited in an experimental technique called **low-energy electron diffraction** (**LEED**), which allows the structure of surfaces to be determined.

8.1 The LEED experiment

In the LEED experiment, as in diffraction techniques that utilize electromagnetic radiation (such as X-ray crystallography), a monochromatic source is required. For an electron beam this means that each electron has the same kinetic energy. In practice, the electrons are emitted by a heated metal or oxide filament. These electrons are then accelerated by an electric potential, which is usually in the range 10–150 eV. Such energies are quite low compared with those used in many types of experiment with electron beams up to, for example, 1 MeV in electron microscopes. This is why the technique is called *low*-energy electron diffraction. More importantly, it means that the incident beam will not penetrate more than a few atom layers into the solid. Hence, LEED examines the structure of the surface alone.

The electron beam is focused, using electrostatic lenses, onto the surface of a single-crystal sample mounted at right-angles to the beam (normal incidence) as shown in Figure 53. Electrons are scattered from the crystal in all directions, including back-scattering in the general direction back towards the source, where they can be observed on a screen. Most of the electrons are scattered inelastically; that is, they lose energy to the surface atoms during the scattering process. These electrons are of no interest in the LEED experiment, so they are prevented from reaching the screen by a negatively charged grid G_1, which repels all electrons with energy less than that of the electrons in the incident beam. *It is the elastically scattered electrons (those that do not lose energy in the scattering process) that contain the diffraction information.* These are able to pass through the grid G_1 before passing a second (earthed) grid G_2 to arrive at the fluorescent screen, where they may be recorded photographically.

As with most surface techniques, it is necessary to maintain an ultra-high vacuum. In LEED this also ensures that electrons are not further scattered by encounter with gaseous molecules.

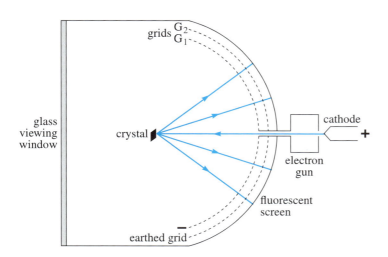

Figure 53 A schematic diagram of the LEED experiment.

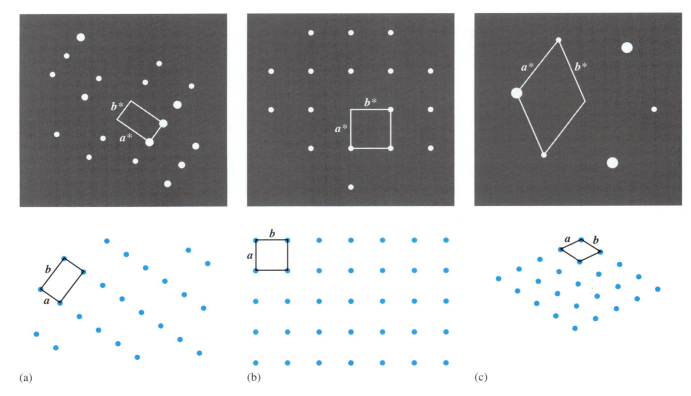

Figure 54 Clean surface LEED patterns (top row) and the corresponding surface structures (bottom row) for: (a) bcc Mo(211), a rectangular mesh; (b) bcc W(100), a square mesh; and (c) fcc Rh(111), a hexagonal mesh. In (a) and (c) the specimen was *not* aligned with the real mesh vectors (*a* and *b*) lying in the conventional directions. Within a LEED pattern the spot intensities differ, a property that is discussed in Section 8.2.3.

8.2 Interpretation of LEED patterns

The LEED experiment produces a pattern consisting of a regular array of spots; some examples are shown in Figure 54. In principle, there should be a spot at the centre of each pattern, but in practice this one is not seen because it coincides with the position of the electron gun (Figure 53), and in any case, it would be obscured by the sample. It is apparent that each of the LEED patterns in Figure 54 can be analysed in terms of a repeating geometrical unit, just as the array of atoms making up a crystal surface can be generated by repetition of the surface unit mesh. For a reason we discuss shortly, the repeating unit in the **LEED pattern** is known as the **reciprocal unit mesh**. The sides of this mesh are identified by two vectors, a^* and b^* ('ay-star' and 'be-star'), with magnitudes a^* and b^*, which correspond to the spacings between the spots in the pattern, as shown in Figure 54. To interpret these patterns, we need to establish the relationships between the magnitudes and directions of a^* and b^*, and those of the surface unit mesh vectors, a and b. The size of an experimental LEED pattern will depend, among other things, on the enlargement involved in recording it photographically. Thus, in Figure 54 the scales for the individual LEED patterns are not the same as those for the corresponding sketches of the surface structures.

8.2.1 The diffraction process

Diffraction of *any* kind is a wave phenomenon, and in the present case depends on the fact that the electrons accelerated in the LEED electron gun to an energy E have a wavelength, λ, given by

$$E = \frac{hc}{\lambda} \tag{12}$$

where h is Planck's constant and c is the velocity of light. First consider diffraction by just one of the rows of atoms making up a surface structure. In the usual experimental arrangement, the incident beam approaches along the surface normal (at right-angles to the surface of the crystal). We can regard it as a collection of individual waves which are exactly 'in phase' with each other; that is, their peaks and troughs are exactly aligned (Figure 55a). On striking the row of atoms, scattering occurs in all directions, in some of which the waves scattered by adjacent atoms will be exactly in phase, and so will interfere *constructively* (reinforce each other — Figure 55a). The net result is a maximum in scattered intensity in these directions. In some other directions, however, the individual scattered waves are exactly 'out of phase': the troughs in one exactly cancel the peaks in another, and the net intensity in these directions is zero (*destructive* interference, Figure 55b).

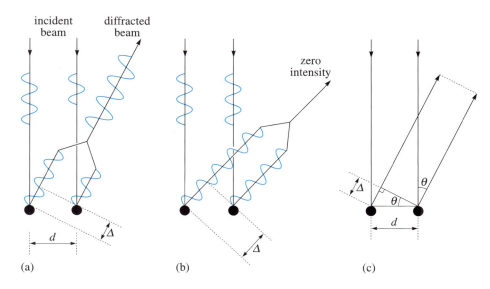

Figure 55 Interference between waves scattered by two surface atoms with spacing d: in (a) the interference is constructive and in (b) destructive; (c) illustrates the relationship between the path difference, Δ, the diffraction angle, θ, and d.

The condition for constructive interference is that the paths travelled by waves scattered by adjacent atoms should differ in length by an amount, Δ, which is a whole number of wavelengths:

$$\Delta = n\lambda \qquad n = 0, 1, 2... \qquad (13)$$

When the angle of incidence is 90°, Figure 55c shows that Δ is also given by

$$\Delta = d \sin \theta \qquad (14)$$

where θ is the angle of diffraction. Combining equations 13 and 14 leads to the familiar expression for the diffraction angles, θ_n, which define the directions of the various diffracted beams:

$$n\lambda = d \sin \theta_n \qquad n = 0, 1, 2... \qquad (15)$$

When a beam strikes the screen, a spot is produced. Although the screen is curved, being part of an imaginary sphere of radius r, with the sample at its centre, what we see through the viewing window (Figure 53) is a projection of the screen image onto a flat plane (e.g. XY in Figure 56). If the observed distance to a particular spot from the centre of the pattern is d_n^*, then Figure 56 shows that

$$d_n^* = r \sin \theta_n \qquad (16)$$

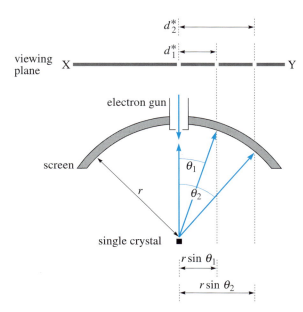

Figure 56 Relationship between diffraction angles and spot positions in a pattern viewed as a projection onto a plane XY.

Then, combining equations 15 and 16, gives

$$d_n^* = \frac{nr\lambda}{d} \tag{17}$$

The first conclusion to be drawn from equation 17 is that all pairs of adjacent spots are equally spaced. Taking, for example, the 'zero order' spot at the centre of the pattern (for which $n = 0$) and the 'first order' spot ($n = 1$), we find that this spacing, d^*, is

$$d^* = \frac{r\lambda}{d} \tag{18}$$

In LEED the screen radius, r, is fixed, and so for a given electron wavelength, λ, d^* is simply proportional to $1/d$. This second conclusion is of central importance.

> Distances, d^*, in any diffraction pattern are reciprocally related to distances, d, in the structure under examination.

The pattern gives us, in fact, a picture of what is known as the 'reciprocal space' of the surface structure (which itself is called 'real space') — hence our earlier reference to the 'reciprocal unit mesh'.

We now move on from a one-dimensional row of atoms to a two-dimensional surface. As we know, its structure is defined by a unit mesh with sides of lengths a and b, and, as we have just seen, the corresponding sides of the reciprocal unit mesh will be in inverse proportion:

$$a^* \propto \frac{1}{a} \quad \text{and} \quad b^* \propto \frac{1}{b} \tag{19}$$

In practice, we are interested in the relative, rather than the absolute, values of a^* and b^*, so the two relationships above are combined to give

$$\frac{a^*}{b^*} = \frac{b}{a} \tag{20}$$

Equation 20 provides one of the two pieces of information necessary for pattern interpretation: **reciprocal mesh vector** magnitudes. The other data required are the directions of a^* and b^*, but here a detailed discussion is beyond the scope of this Course, so we ask you to accept the following:

> a^* is perpendicular to b b^* is perpendicular to a (21)

8.2.2 LEED pattern spot positions

We are now in a position to re-examine the three LEED patterns in Figure 54. In the case of molybdenum, a *bcc* metal, the real mesh on the (211) surface is rectangular (see Figure 44). With this mesh orientated in the *conventional* way (Figure 57a), a is perpendicular to b, and so a^* lies in the same direction as a. Similarly, b^* lies in the same direction as b (relationships 21). Because in real space a is *smaller* than b, the situation in reciprocal space is reversed, with a^* being the *longer* of the two vectors (equation 20). The net effect is that the reciprocal mesh is rectangular, but it appears to be rotated 90° with respect to the real mesh (and the vector labels are interchanged).

When the real mesh is square, as it is for the (100) surface of the *bcc* metal, tungsten (Figure 57b), the situation is even simpler. The reciprocal mesh is also square, and conversion between pattern and structure involves only an interchange of vector labels. The third case, however — a hexagonal mesh (Figure 57c) — illustrates the full implications of relationships 21. Here, a is *not* perpendicular to b and so the directions of a^* and a (or of b and b^*) no longer coincide.

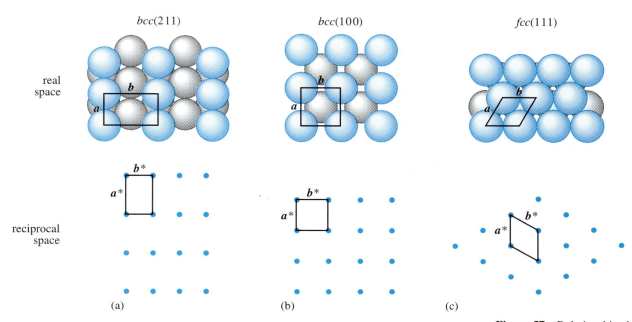

Figure 57 Relationships between three simple types of real and reciprocal mesh. Comparison with Figure 54 reinforces the point that in Figures 54a and c the specimen was not aligned with the real mesh vectors lying in the conventional directions.

Note that the lengths used for the reciprocal vectors are unimportant, provided that the ratio, a^*/b^*, is correct. As indicated earlier, the size of an experimental LEED pattern will depend, among other things, on the enlargement involved in recording it photographically.

In the most general case the surface unit mesh is *oblique*, having sides of unequal length and a mesh angle with some value other than 90° or 120°. Prediction of the corresponding LEED pattern then involves the steps listed in Box 3 (overleaf); these are also illustrated in Figure 58.

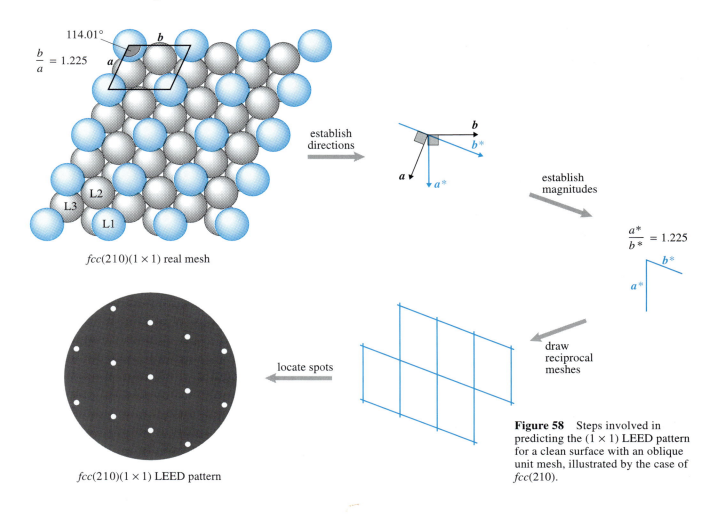

Figure 58 Steps involved in predicting the (1×1) LEED pattern for a clean surface with an oblique unit mesh, illustrated by the case of $fcc(210)$.

> **Box 3   Predicting the LEED pattern for a clean-surface structure**
>
> (i) Draw the real unit mesh in the conventional way.
>
> (ii) Establish the *directions* of the reciprocal mesh vectors (relationships 21).
>
> (iii) Establish the *relative magnitudes* of the reciprocal mesh vectors (equation 20).
>
> (iv) Draw a reciprocal mesh and then duplicate it to build up an area of reciprocal space.
>
> (v) Draw a 'spot' at the corners of the meshes to generate the 'pattern'.

The structures of most crystals studied in LEED experiments, usually metals or metal oxides, are known. So, when dealing with a clean surface, its structure is also known and its diffraction pattern can be worked out in the way described above. We are then interested in any changes that may be observed as a result of adsorption. Provided the adsorbate layer is ordered, we saw in Section 7.2 that it can be described in terms of an $(m \times n)$ unit mesh. Because, virtually always, $m \geqslant 1$ and $n \geqslant 1$, the adsorbate real-space mesh is either the same size as, or more often *bigger* than, the clean surface (1×1) mesh. It then follows that the *reciprocal* mesh of the adsorbate structure may be the same size as that of the 'clean' pattern, but is more likely to be *smaller*, by factors of $1/m$ in the a^* direction and $1/n$ in the b^* direction. And smaller reciprocal meshes mean more spots.

The LEED pattern for an $(m \times n)$ layer can be predicted in a similar way to that used for the clean surface. Again, the steps involved are listed in Box 4, and the procedure is illustrated in Figure 59.

> **Box 4   Predicting the LEED pattern for an adsorbate-covered surface**
>
> (i) Identify the (1×1) mesh of the clean substrate and the $(m \times n)$ mesh of the adsorbate structure.
>
> (ii) Establish the (1×1) reciprocal unit mesh vectors a^* and b^* (using the procedure illustrated in Figure 58).
>
> (iii) Establish the magnitudes of the $(m \times n)$ reciprocal mesh vectors as fractions $1/m$ and $1/n$ of a^* and b^*.
>
> (iv) Draw an adsorbate reciprocal mesh and then duplicate it to build up an area of reciprocal space.
>
> (v) Draw a 'spot' at the corners of the meshes to generate the $(m \times n)$ 'pattern'.

There is an important point to note about the LEED 'pattern' in both Figure 58 and in Figure 59. All spots appear to have the same intensity, but comparison with actual patterns, Figure 54 for example, shows that this is incorrect. There will always be intensity differences between various spots, an issue to which we shall return briefly in Section 8.2.3.

When there is no rotation between the adsorbed layer and the substrate, Figure 59 shows that between each spot position in the 'clean' pattern, adsorption introduces $(m - 1)$ rows of extra spots in the a^* direction and $(n - 1)$ extra rows in the b^* direction. (Of course, if the adsorbed layer has a (1×1) structure, no additional spots appear.) An actual LEED example is shown in Figure 60, where adsorption produces a (2×2) mesh, giving one additional row of spots in both a^* and b^* directions (Figure 60b).

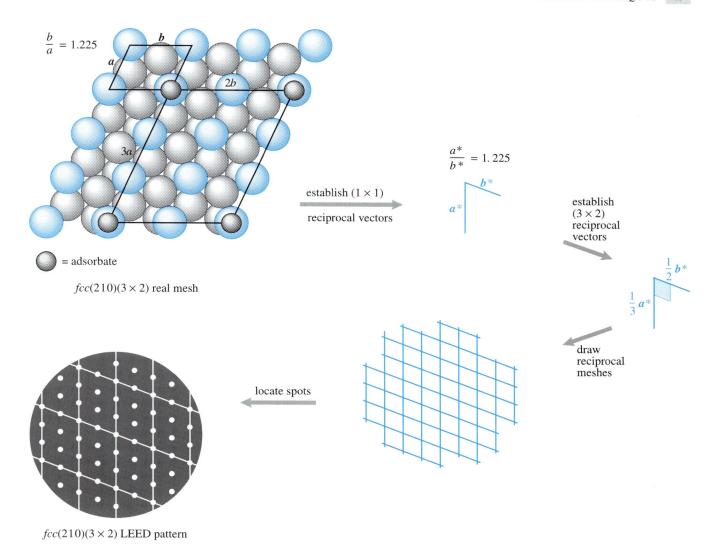

Figure 59 Steps involved in predicting the LEED pattern for an adsorbate-covered surface with an oblique unit mesh, illustrated by the case of an $fcc(210)(3 \times 2)$ layer. The reciprocal mesh of the clean surface is indicated as fine lines in the LEED pattern.

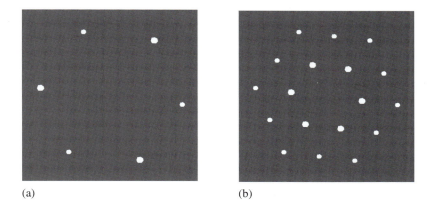

Figure 60 The change in a LEED pattern caused by adsorption: (a) a clean Rh(111) surface; (b) the same surface with a low CO coverage.

If there *is* a rotation between the adsorbed layer and the substrate as, for example, in Figure 49, the procedure illustrated in Figure 59 can still be used to derive the LEED pattern, provided the appropriate rotation is introduced when establishing the adsorbate reciprocal mesh vectors; an example is provided by SAQ 15 (Section 8.3).

In practice, we would normally apply the procedure outlined in Figure 59 in *reverse*, using an experimental pattern to determine the dimensions of the adsorbate mesh in *real* space. This is equivalent to working 'backwards' through Figure 59.

STUDY COMMENT The following SAQ provides an opportunity to practise this yourself. You should try it before proceeding further.

SAQ 14 Platinum is a *fcc* metal and the Pt(410) surface has a terraced structure. The LEED pattern of a (3 × 1) adsorbate layer on such a surface is shown schematically in Figure 61. Assuming that the sample has been orientated in the conventional way, deduce the directions and relative magnitudes of the vectors, ***a*** and ***b***, defining the (1 × 1) clean-surface unit mesh in real space. [*Hint* Start by establishing the ***a**** and ***b**** directions, and hence identifying the (3 × 1) reciprocal mesh in the LEED pattern.]

8.2.3 LEED pattern spot intensities

Up to this point we have assumed that the real-space surface structure is *two-dimensional*, described by a *unit mesh*, and this view has been adequate because our attention has been restricted to the *positions* of spots in the pattern.

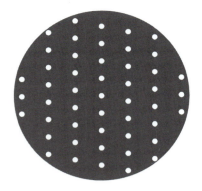

Figure 61 The Pt(410)(3 × 1) LEED pattern.

> The size and shape of the reciprocal unit mesh (i.e. *spot positions*) depend only on the size and shape of the real-space unit mesh.

In Section 7, however, we saw that 'surface' structure is described more properly by a *three-dimensional* 'surface unit cell', extending one or more atomic layers into the solid beneath the unit mesh. One example of the significance of this concept was given in Figure 51, which illustrated three different adsorption sites on a particular surface being occupied by a particular adsorbate. In each case the unit *mesh* was (2 × 2) and so the spots in the corresponding LEED patterns would all be in the same *positions*. The surface unit *cells*, however, differed in the locations of the metal atoms in relation to the adsorbate species, and these differences would be reflected in the *intensities* of the diffraction spots. Any given spot would be found to have a different intensity in each of the three patterns. This is an example of a second important principle.

> LEED *spot intensities* depend on the locations of individual atoms within the surface unit cell.

So, determination of the fine details of a surface structure by LEED requires an analysis of spot intensities. The theory involved is quite complex, and satisfactory computational methods became available only around 1980. Now, however, intensities are recorded by video camera, digitized, and fed directly into a computer, making the analysis a fairly routine matter, at least for relatively simple structures.

8.3 Summary of Section 8

1 In low-energy electron diffraction, LEED, a beam of monochromatic electrons is diffracted by the surface of a single crystal.

2 Diffraction by the regular array of atoms on the surface (the real unit mesh) produces a regular array of spots in the LEED pattern (the reciprocal unit mesh).

3 In the diffraction process, the distance, d^*, in the diffraction pattern is reciprocally related to the distance d in the surface structure.

4 The magnitudes of the surface unit mesh vectors (***a*** and ***b***) are related to the magnitudes of the reciprocal unit mesh vectors (***a**** and ***b****) as:

$$\frac{a^*}{b^*} = \frac{b}{a}$$

5 The directions of the unit mesh vectors are related to the directions of the reciprocal unit mesh vectors as:

   a* is perpendicular to ***b***; ***b**** is perpendicular to ***a***

6 The spot positions in a LEED pattern depend on the size and shape of the real-space unit mesh.

7 The LEED spot intensities depend on the locations of individual atoms within the surface unit cell.

8 LEED results are interpreted by using the above relationships to establish the real unit mesh (surface structure) from the reciprocal unit mesh (LEED pattern). This may be done for a clean single crystal surface or for an ordered layer of adsorbate on such a surface.

> **STUDY COMMENT**   The best way to ensure that you have grasped the main points in this Section, and that you can apply the procedures in Boxes 3 and 4, is to work through SAQ 14 (Section 8.2.2) and SAQs 15 and 16 (below). You should do this now before proceeding to Section 9.

SAQ 15   Using the procedure described above and illustrated in Figure 59, derive the LEED pattern for a $(\sqrt{2} \times \sqrt{2})R45°$ adsorbed layer on an $fcc(100)$ surface. [*Hint*   When you have established the magnitudes of the $(\sqrt{2} \times \sqrt{2})$ reciprocal mesh vectors (step (iii) in Box 4), you should rotate these vectors *clockwise* by the rotation angle *before* drawing the reciprocal mesh.]

SAQ 16   Figure 62 shows LEED patterns for (a) a clean low-index surface of tungsten (a *bcc* metal) and (b) the same surface after exposure to oxygen and heating. What can you deduce about the structures of the clean and oxidized surfaces? (When comparing the two patterns, you should allow for the fact that (a) was recorded at a slightly lower photographic magnification than (b)).

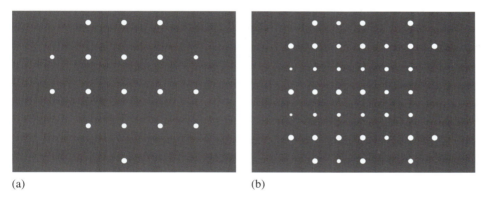

(a)   (b)

Figure 62   (a) The LEED pattern of a clean low-index surface of tungsten; (b) the LEED pattern of the same surface after exposure to oxygen and heating.

9 VIBRATIONAL SPECTROSCOPY

For the purposes of **vibrational spectroscopy**, molecules are envisaged as a collection of atoms, vibrating with frequencies characterized by their masses and the force constants of the bonds (analogous to springs) that connect them. In the study of bulk phases, two techniques are employed. In infrared spectroscopy, vibrational motion in molecules is excited by the absorption of infrared radiation. In Raman spectroscopy, scattered radiation is changed in frequency as it either gains or loses energy to molecular vibrations. The two techniques are governed by different selection rules which determine the molecular vibrations that may be studied, and so they are complementary.

As you may recall from the Second Level Inorganic Course, for diatomic molecules, the vibrational frequency v leads directly to a value of the force constant k for the bond in the molecule according to equation 22

$$v = \frac{1}{2\pi}\left(\frac{k}{\mu}\right)^{1/2} \tag{22}$$

Here μ is the reduced mass of the diatomic molecule AB:

$$\mu_{AB} = \frac{m_A m_B}{m_A + m_B} \tag{23}$$

For somewhat larger molecules, vibrational spectroscopy enables the symmetry and, in favourable cases, the structure of a molecule to be deduced. Even larger molecules are considered as a collection of, often independently vibrating, fragments or groups with characteristic vibrational frequencies. Recognition of such group frequencies provides an empirical means of identifying groups of atoms within a molecule.

These same properties may be elucidated in the study of the vibrational spectroscopy of surfaces: identification of groups (or small molecules) by group frequencies, estimation of force constants, establishment of symmetry. In surface chemistry, vibrational spectroscopy is therefore applied very largely to the study of *adsorbates* — their identification, bonding and alignment to the surface.

For adsorbates on single crystal surfaces, there are various ways in which vibrational information can be obtained, and we shall consider two of the most widely applicable, which are based on very different principles. The first of these, exactly analogous to infrared spectroscopy, is **infrared reflection absorption spectroscopy**. The second is the study of electrons that are scattered *inelastically* when a beam of electrons is directed at a surface (in contrast to LEED, Section 8, which involves the *elastically* scattered component). The inelastic scattering results from the energy loss caused by exciting the vibration of surface groups.

9.1 Infrared Reflection Absorption Spectroscopy (IRAS)

In the IRAS experiment the absorption is measured after the incident infrared beam is reflected from the surface, usually at a high angle of incidence, θ_i (see Figure 63), which, for reasons associated with the basic physics of the process, enhances the absorption. It is a common observation that only those vibrations that have an oscillating dipole, or some component of it, *perpendicular* to the surface (the z direction in Figure 63) are active in IRAS. There are theoretical reasons to suppose that 'perpendicular' absorptions are much more intense than those involving vibrations parallel to the surface. This tendency for selective absorption has become known as the *surface selection rule*. The rule is not rigid, but it does provide useful empirical guidance in the spectral range commonly covered by IRAS.

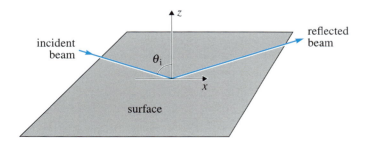

Figure 63 The experimental arrangement for IRAS.

As IRAS is just one technique for studying molecular vibrations in adsorbates, it is useful for comparative reasons to consider its performance. Like many modern forms of spectroscopy, it gains from use in a Fourier transform mode, such as was discussed in the context of infrared spectroscopy in the Second Level Inorganic Course. The details of the method are unimportant here; the technique is a pulsed method as opposed to continuous irradiation, and it enhances both sensitivity and resolution. The sensitivity of current instruments is about 10^{-3} of a monolayer coverage for adsorbates that absorb strongly. The commonly studied and important species CO is one of these.

With readily available spectrometers, the range of frequency (or more strictly wavenumber)* is about $1\,000 - 4\,000\,\text{cm}^{-1}$ although with special sources of radiation this can be extended to about $400\,\text{cm}^{-1}$. High-quality Fourier transform instruments give a resolution of about $1\,\text{cm}^{-1}$, which is superior to the requirements of the method, because the inherent width of absorptions of surface species is usually more than $10\,\text{cm}^{-1}$. As the technique employs only electromagnetic radiation rather than electron beams, it does not require low pressures, and can therefore be used under conditions more closely resembling those in catalysis.

9.2 Electron Energy Loss Spectroscopy (EELS)

When the surface of a solid is irradiated with a beam of electrons, some are scattered elastically (as in LEED) and some inelastically. Inelastic scattering results from various processes including energy transfer to the solid, for example, to excite electronic transitions as in Auger spectroscopy. Here we are concerned with the excitation of the *vibrations* of surface species, which, because of the resolution needed to observe vibrational transitions, is sometimes called **high-resolution electron energy loss spectroscopy**, **HREELS**, or often just referred to as **electron energy loss spectroscopy**, **EELS**. Here we shall simply use the acronym EELS.

In EELS, a monochromatic electron beam (fixed energy) with energy up to 300 eV is directed at the surface of a single crystal, as in Figure 64, and the scattered beam passes into an energy analyser similar to that used in PES. Most of the beam is scattered randomly (diffuse scattering). Only a few per cent of the electrons are scattered in what is called the *specular*† direction (where the scattering angle, θ_s, is equal to the incident angle, θ_i, Figure 64). Clearly, a number of features of the experiment may be varied — the energy of the electron beam and the angle at which scattering is observed, which can give structural information. But in most experiments the electrons are observed in the specular direction because in this direction the inelastic scattering (or absorption) is more intense than other directions.

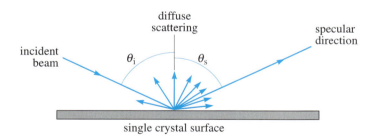

Figure 64 Electron scattering in the EELS experiment.

---

* Note that as in all forms of infrared spectroscopy it is conventional to give the 'frequency' in units of cm^{-1}, which are not really frequency units, but wavenumber, $\bar{\nu}$, or reciprocal wavelength. Wavenumber and frequency are proportional, being related as

$$\bar{\nu} = \frac{\nu}{c}$$

where c is the speed of light.

† From the Latin *speculum*, meaning mirror.

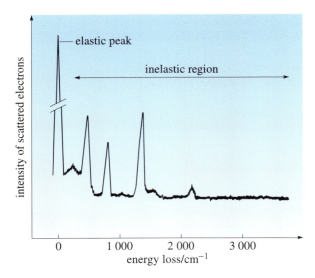

Figure 65 An EELS spectrum, showing the inelastic region.

Figure 65 shows a typical spectrum, a plot of intensity of scattered electrons versus energy loss, expressed in either meV or cm^{-1} (1 meV ≈ 8 cm^{-1}). The top of the intense elastic peak is much scaled down in this spectrum; each of the other, very much smaller, peaks corresponds to the excitation of a particular vibration, usually due to an adsorbate. Some aspects of EELS are evident from Figure 65.

■ How do the frequency range and resolution compare with those of infrared spectroscopy, IRAS?

■ In Figure 65 the first peak in the inelastic region is evident at about 250 cm^{-1}, and so the accessible range of frequency extends to lower values than in IRAS. On the other hand, the resolution of this spectrum is poor: peak widths (usually measured half-way up the peak) are more than 50 cm^{-1}, which is much larger than in infrared spectroscopy. This is largely due to the limitations imposed by the resolution of the spectrometer.

As with IRAS, the familiar relationship (equation 22) between the absorption frequency and the properties of a vibrating molecule applies.

■ In which region of a vibrational spectrum would you expect to observe the absorption due to an adsorbate atom such as carbon or oxygen bound to a surface metal atom such as copper (that is, a Cu—C or Cu—O bond)?

■ This absorption will be at low frequency, and hence low wavenumber, due to the large value of the reduced mass, μ, that results from the heavy metal atom (equation 23).

The vibrations of adsorbate–surface bonds, typically light atom–metal atom bonds, generally lie well below 1 000 cm^{-1}, in what is called the **far infrared**. So EELS is particularly well suited to the study of adsorbate–substrate interactions, because of its wide spectral range.

Due to the presence of many neighbouring surface atoms, the effect of the surface on adsorbates is usually to broaden the vibrational absorption band, so that the relatively poor resolution of EELS is not always a great disadvantage. In fact, this is overcome in some modern instruments, in which improved electron optics allow a resolution better than 10 cm^{-1}.

As with other techniques that employ electron beams, EELS necessitates the use of low pressures, a condition not imposed by infrared spectroscopy.

■ How does the use of low pressure affect the application of EELS to the study of catalysis?

■ In practice, catalysts are used at atmospheric or, commonly, much higher pressures. So EELS studies give information under very different conditions.

In another respect, EELS resembles infrared spectroscopy. In IRAS a selection rule operates which requires that an active vibration must have its changing dipole (or at least some component of it) aligned perpendicular to the surface. In EELS the same rule applies for scattering observed in the specular direction (the usual mode of operation), but not necessarily if observations are made in other scattering directions.

9.3 Application of IRAS and EELS

For reasons of ease of use and wide-ranging application, vibrational spectroscopy is the technique of choice for identifying groups of atoms, and examining their symmetry, in most chemical studies of bulk samples. The same applies in surface studies. The requirement of the oscillating dipole has focused much attention on the adsorption of CO, usually on metals.

Although the interpretation of vibrational spectroscopic studies considers changes in bond strength (or force constant) and the general use of equation 22, it also employs the empirical approach. For this purpose, model molecular compounds are needed in which the adsorbate of interest, for example CO, is bound in different ways to metal atoms. These model compounds, often involving small clusters of metal atoms, must have known structures, which are usually deduced by X-ray diffraction.

For carbon monoxide, the models are metal carbonyls in which three types of metal–CO bonds are observed. These are illustrated in Figure 66, which shows examples of linear bonding (known as Type II) and threefold bridge (known as Type III), together with one involving CO ligands in both linear and twofold bridging (known as Type I) positions. Spectroscopically, these are distinguished by the CO stretching frequencies indicated in Figure 66. Type I has two frequencies, since it contains two types of CO, whereas Types II and III each has a single absorption.

As an example of the application of this approach, an EELS study of the non-dissociative adsorption of CO on the Rh(111) surface gives the following results. At a very low coverage of CO, an absorption frequency is observed in the EELS spectrum at about $2\,020\,cm^{-1}$. As coverage by CO is increased, an additional band at about $1\,850\,cm^{-1}$ appears. A further band is present at about $460\,cm^{-1}$ at all coverages (see SAQ 18 in Section 9.4).

■ On the basis of the vibrational absorptions of the model compounds in Figure 66, to which adsorption sites would you assign the two high-frequency bands described for CO on Rh(111)?

□ Comparison with the data in Figure 66 suggests that frequencies at or above $2\,000\,cm^{-1}$ are characteristic of linearly bonded CO. So at very low coverage, all the adsorbed molecules appear to be located directly on top of substrate atoms, in terminal adsorption sites. The appearance of the band at $1\,850\,cm^{-1}$ indicates the additional occupation of twofold bridge sites at higher coverage.

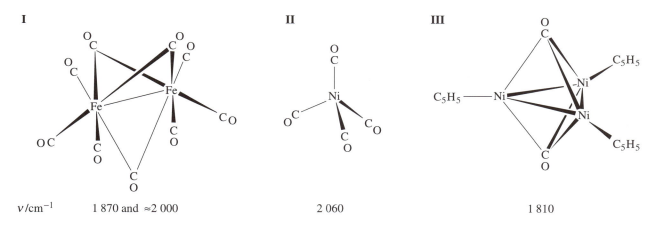

Figure 66 Types of CO clusters and their C—O vibrational frequencies.

The dependence of the CO stretching frequency on binding geometry can be understood in terms of the charge transfer that takes place between CO and the surface during adsorption. Within carbon monoxide, the carbon–oxygen bond is described by the molecular orbital diagram introduced in Section 5 (Figure 12). The orbitals of note in forming bonds from CO to other atoms are the highest occupied molecular orbital (HOMO, labelled 5σ) and the lowest unoccupied molecular orbital (LUMO, labelled $2\pi^*$). Two types of bonding can occur between CO and the metal. In the first type, the 5σ HOMO donates charge to the metal conduction band, forming a σ bond from carbon to rhodium. The second contribution to the Rh—C bond comes from a π interaction involving back-donation of charge from the metal conduction band into the LUMO. The extent of this transfer by back-bonding increases with the number of rhodium atoms to which a CO molecule is attached.

■ What would you predict to be the effect on the force constant of the carbon–oxygen bond, its force constant and its stretching frequency when electronic charge is accepted by CO in π back-bonding?

▪ In back-bonding, the charge is transferred into the $2\pi^*$, *antibonding* orbital. This decreases the bond strength and the force constant. As a result, the vibrational frequency is decreased (equation 22).

This expectation is borne out by the shift of the absorption frequency to lower values as the number of surface Rh atoms to which each CO is coordinated increases. Similar results to those of IRAS have been obtained in EELS studies of the same system.

The interpretation exemplified by CO adsorption modes does not unfortunately extend to other systems such as the comparison of cluster and surface site spectroscopy of nitric oxide, NO. Even with CO a word of caution is necessary because recent studies by LEED and IRAS reached different conclusions regarding the bridged sites of CO on a Ni(111) surface.

Most catalysts involve porous oxides, either as the active component or as a support for a highly dispersed metal. These materials cannot be examined by EELS because electrons cannot penetrate the solid to reach the surfaces within the pores, which constitute a very large part of the total surface area. Nor can they be examined by IRAS, which relies on radiation striking a *flat* metal surface at a high angle of incidence. On the other hand, the oxides Al_2O_3 and SiO_2 are transparent to infrared radiation. Therefore, supported metal catalysts can be studied using *standard* infrared techniques. So there are a number of ways in which infrared spectroscopy finds application in catalytic studies.

9.4 Summary of Section 9

1 Two techniques exist for the study of vibrational spectra of adsorbates. In infrared absorption spectroscopy, IRAS, infrared radiation is used. In electron energy loss spectroscopy, EELS or more correctly HREELS, a beam of monochromatic electrons is used.

2 In IRAS, the absorption of infrared radiation excites molecular vibrations in adsorbates. It may be used at atmospheric pressure, or higher.

3 A surface selection rule operates in IRAS: vibrational modes perpendicular to the surface give the most intense absorptions.

4 In EELS, electrons are scattered inelastically, losing energy to excite molecular vibrations.

5 A perpendicular selection rule operates in EELS in the specular direction.

6 Vibrational studies of surface adsorption of CO may be related to the coordination of CO in small clusters of metal atoms.

SAQ 17 List the relative advantages and disadvantages of IRAS and EELS as methods for studying vibrational modes in adsorbates.

SAQ 18 In the EELS spectrum of CO on a Rh(111) surface, a band appears at 460 cm^{-1}. What is the likely vibration that causes this absorption?

SAQ 19 When N_2 is adsorbed on the Fe(111) surface at 110 K the EELS spectrum observed is that in Figure 67a. When the sample is heated to 170 K, the spectrum changes to that in Figure 67b. The gas-phase frequencies of nitrogen–nitrogen stretching modes are: N≡N, 2 331 cm^{-1}; N=N, 1 530 cm^{-1}; N—N, 1 098 cm^{-1}. Using this information together with equation 22, suggest an interpretation of the EELS spectra.

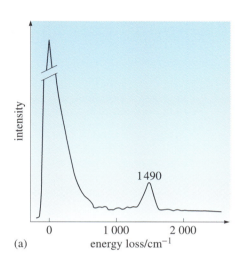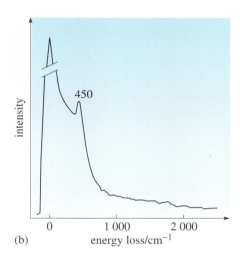

Figure 67 EELS spectra of nitrogen gas on Fe(111): (a) the spectrum when nitrogen is adsorbed at 110 K; (b) the spectrum when the same sample is heated to 170 K.

10 X-RAY ABSORPTION SPECTROSCOPY

X-ray spectroscopy has been studied since the 1920s, but recently it has been put to new use for both bulk and surface investigations. The reasons for renewed interest are the development of the powerful *tunable* X-ray source, the **synchrotron**, and a theoretical understanding of the factors affecting X-ray absorption. When a solid is exposed to an X-ray beam, the X-rays are absorbed over a wide range of energy, with peaks resulting from excitation and ionization. As you know from Section 5, the study of the emitted photoelectrons is the basis of the XPS technique. In this Section we examine how the intensity of the X-ray beam is itself affected, and how this phenomenon is interpreted.

Typically, an **X-ray absorption spectrum** has the form shown schematically in Figure 68. The spectrum can be divided into three regions, each of which is accounted for by a different process. At low energy lies a region in which absorption decreases with increasing photon energy. This (pre-edge) region is the smooth background that results from the scattering of X-rays by matter.

Against this background, the structure of the spectrum is a sequence of steep discontinuities, one of which is shown in Figure 68. These occur at energies that

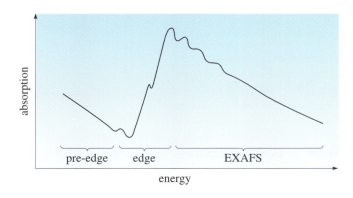

Figure 68 A typical X-ray absorption spectrum.

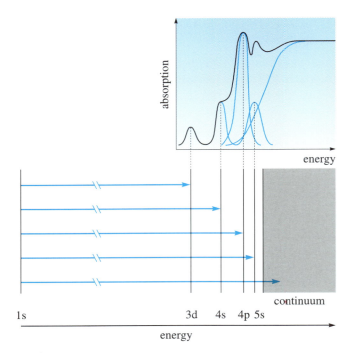

Figure 69 Transitions that give rise to structure in the edge region, corresponding to ionization of a 1s electron. The separate absorption processes that contribute to the overall absorption curve are shown in blue.

are characteristic of the element that absorbs the X-rays, and are called 'edges'. As Figure 69 shows, these edges have fine structure, which can be explained as the overlap of a set of absorption peaks and the onset of ionization. These absorptions are the result of transitions of a core electron to high energy levels and into the continuum, this last process being the ionization. The relationship between the transitions and a typical spectrum is illustrated in Figure 69.

The third spectral region lies at photon energies that range from about 40 eV above the edge to high energies. Again, as Figure 68 shows, this region contains fine structure; the interpretation of this fine structure is the major focus of this Section.

10.1 Extended X-ray Absorption Fine Structure (EXAFS and SEXAFS)

STUDY COMMENT The events that give rise to EXAFS are shown in the video sequence (band 7, *A clean getaway*, on videocassette 2) associated with Topic Study 2. You may wish to come back to this Section after you have viewed that sequence.

The successful interpretation of the **extended X-ray absorption fine structure**, **EXAFS**, was developed in the 1970s and is now used as a structural tool for the investigation of a diverse range of materials. The fine structure, or oscillation, of the spectral intensity in the EXAFS region is known to be the result of interaction between the emitted photoelectron and neighbouring atoms. If this process is considered as a sequence of events, it is inexplicable that an interaction *following* emission of the electron should affect the absorption of the *incoming* X-ray photon that produced the photoemission. To understand the EXAFS process, the absorption, photoemission and interaction must be envisaged as *simultaneous*, and the photoelectron viewed as a wave. You can think of this wave as a ripple emanating from the atom (Figure 70a), although it is, of course, propagated in three dimensions.

In such a model, the wave is back-scattered by neighbouring atoms, and interferes with the outgoing wave at the site of photoemission. Two cases of interference are shown, one in which the back-scattered wave interferes constructively (Figure 70b) and so enhances the absorption, and the other at a different photon energy, such that the scattered wave interferes destructively (Figure 70c), causing a minimum in the absorption. As in low-energy electron diffraction, the nature of the interference depends on the electron wavelength and a path difference, between the outgoing and back-scattered waves in EXAFS. In this case, the path difference is determined by the

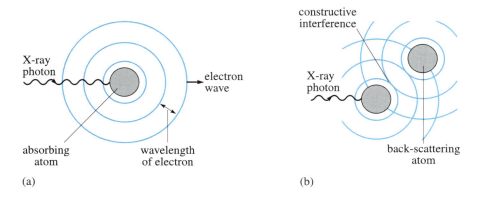
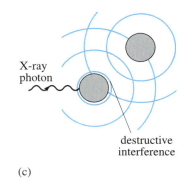

distance between the emitting and scattering atoms, while the electron energy and hence wavelength (equation 12), varies with changing X-ray energy. So, for a particular interatomic distance, the conditions for constructive and destructive interference (equation 13) are successively satisfied as the X-ray spectrum is scanned in wavelength. The resulting maxima and minima occur in the absorption due to the EXAFS process, seen in the spectrum as oscillations against the background (Figure 68).

Figure 70 The EXAFS process: (a) the generation of an electron wave by ionization; (b) constructive interference in scattering of the electron wave; (c) destructive interference.

The scattering process pictured in Figure 70 is an obvious oversimplification. The wave representing the emitted electron is scattered not only by the *nearest* neighbours, but by all neighbouring atoms. As you might expect, the extent of the interference due to this back-scattering depends on several factors. One, as we have seen, is the distance of the neighbouring atom from the absorbing atom, which is the photoelectron source. Another is any phase change that the electron wave may undergo on being scattered. The third is the scattering ability of the neighbouring atoms. As the scattering of X-rays is caused by the electrons in an atom, this second factor is determined by the number of neighbouring atoms and their atomic numbers.

As with most experimental techniques, it is the factors that govern the phenomenon that may be deduced from its study. So from EXAFS we expect to determine the distance from the photoelectron source to each neighbouring atom, and the number and type of neighbouring atoms. Small differences in atomic number are difficult to discern since they only have a small effect on the EXAFS signal. So, though it is not easy to distinguish between elements adjacent in the Periodic Table, such as nitrogen and oxygen, it is usually possible to distinguish members of the same Group, such as oxygen and sulphur. In the light of other chemical information about the sample under study, this limitation is not often a disadvantage.

The interpretation of EXAFS can be illustrated with the simple example of germanium, an element that crystallizes with the diamond structure, Figure 71a. An EXFAS spectrum for a powder sample of germanium is shown in Figure 71b. The procedure by which the interatomic distances are extracted from such data involves a stepwise mathematical treatment, which we need not consider. The result is a curve of the type shown in Figure 71c, known as the '**radial distribution function**', which is an oscillating function of internuclear distance, R, from the absorbing atom. It has peak

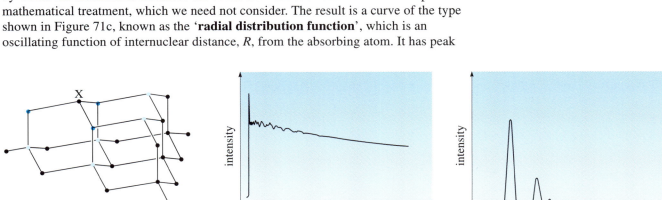

Figure 71 EXAFS of germanium: (a) the diamond-like structure of germanium, showing the first coordination shell of atoms (nearest neighbours of atom X) in dark blue and the second coordination shell (next nearest neighbours) in light blue; (b) the EXAFS spectrum of a powder sample of germanium; (c) the corresponding radial distribution function, plotted as intensity against internuclear distance.

positions corresponding to the distances from the absorber to its neighbouring coordination shells, and peak amplitudes related to the number of neighbouring atoms and to the identities of the absorber and its neighbours. Above the general noise level in Figure 71c it is possible to discern that each atom of germanium is surrounded by shells of neighbours, known as **coordination shells**, the first three being at about 220 pm, 370 pm and 440 pm separation.

Because of the penetrating nature of X-rays, EXAFS obviously examines the bulk of a solid, but X-ray crystallography is, in fact, a superior method for structure determination. Nevertheless, EXAFS is of considerable general importance since, unlike X-ray diffraction, it is not restricted to crystalline samples. In addition to powders, as in the example above, it can be used with amorphous materials and even with liquids. A further advantage is that it is possible to obtain separate information on each type of atom in a complex solid, since X-ray absorption occurs at different photon energies for different elements.

To illustrate the use of the technique in catalysis, we consider an EXAFS comparison of two catalysts: 1% Ru dispersed on SiO_2 (99%) and 1% Ru–0.63% Cu (1 : 1 atomic ratio) on SiO_2. The ruthenium–copper system is used for hydrocarbon production from CO and H_2, with the copper added to suppress undesirable hydrogenolysis of products, for example of C_2H_4 to CH_4, which occurs with ruthenium on its own. In the bulk phase, ruthenium and copper are immiscible.

- What do the known properties of the ruthenium–copper and ruthenium catalysts, and of the ruthenium and copper bulk phases, suggest about the structure of the ruthenium–copper catalyst?

- Because of its effect on the catalytic process, copper cannot be present in isolation from ruthenium on the surface of the support; there must be intimate contact between the two metals. Also, because of the immiscibility of the bulk phases it is very unlikely that the small catalyst particles will consist of a 1 : 1 ruthenium : copper alloy. This leaves two possibilities: that ruthenium particles are coated with a thin layer of copper or that copper particles are coated with a thin layer of ruthenium.

EXAFS radial distribution functions for the two catalysts and for a Cu–SiO_2 sample are shown in Figure 72.

- Examine Figure 72 and comment on:

    (a) the environment of a ruthenium atom in the ruthenium–copper catalyst compared with that of a ruthenium atom in Ru–SiO_2 (Figure 72a);

    (b) the environment of a copper atom in the ruthenium–copper catalyst compared with that of a copper atom in Cu–SiO_2 (Figure 72b).

- It is clear from the close similarity of the distribution functions in Figure 72a that the environment of a ruthenium atom is affected very little by introduction of copper into the system. In Figure 72b, however, the two functions differ significantly and so the environment of a copper atom in the ruthenium–copper catalyst is *not* the same as in pure copper particles.

We can conclude that in the ruthenium–copper catalyst the nearest neighbours to a ruthenium atom are mostly other ruthenium atoms, as they must be when ruthenium is present on its own. Thus, in the ruthenium–copper case the ruthenium must exist mainly in the form of small particles. (A thin layer of ruthenium, either covering copper particles or sandwiched between copper and the silica support, is ruled out.)

The opposite is true for copper. Not all the nearest neighbours to a copper atom are other copper atoms. A significant number must be ruthenium atoms. Given that the ruthenium is arranged as small particles, it follows that the copper probably forms a thin coating on their surfaces. This picture is consistent with the observation that addition of even small amounts of copper to ruthenium has a considerable effect on the catalytic properties of the system.

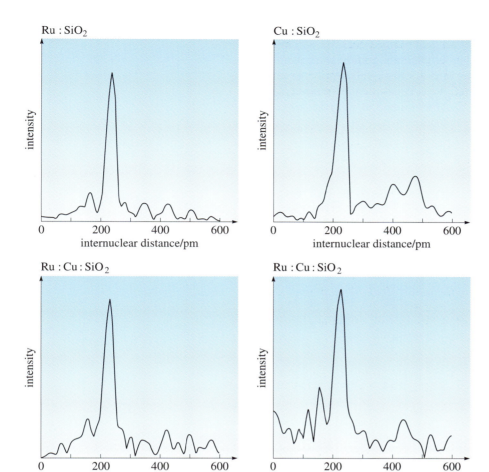

Figure 72 (a) The Ru EXAFS of the ruthenium–copper catalyst (bottom) compared with the EXAFS of supported ruthenium (top); (b) the Cu EXAFS of the ruthenium–copper catalyst (bottom) compared with the EXAFS of supported copper (top).

A more quantitative analysis can be carried out by assuming various structures for the copper-covered ruthenium particles, *calculating* the corresponding EXAFS spectra, and comparing with the experimental data to find the best fit. In this way it is estimated that the first (nearest-neighbour) coordination sphere of a ruthenium atom contains about 11 atoms, of which almost 10, on average, are other ruthenium atoms. In contrast, the first coordination sphere of a copper atom contains 9 atoms, of which 3 are ruthenium.

If we assume that the ruthenium–copper particles are cubic in shape and that the copper layer is one atom thick, we can now estimate the particle size. As the metals are in the atomic ratio Ru : Cu = 1 : 1, it follows that the number of ruthenium atoms forming the core should be approximately equal to the number of copper atoms in the coating. A cube measuring ten atoms along each side would then contain $(8 \times 8 \times 8) = 512$ ruthenium atoms in its core and have $(10 \times 10 \times 10) - 512 = 488$ copper atoms covering its six sides. Smaller cubes have a higher surface : volume ratio, and larger cubes a smaller ratio. This result suggests that the particles are about 1 000 atoms in size.

■ Copper is a *fcc* metal with unit cell parameter $a_{0,Cu} = 361$ pm. Assuming that the arrangement of copper atoms on the faces of the cubic particles is the same as that on a Cu(100) surface (Figure 38), what will be the length of the cube sides?

▪ The nearest-neighbour spacing on the *fcc*(100) surface is $\sqrt{2}a_0/2$ (Figure 38) and there will be nine spacings between the ten atoms forming a cube side. Thus, the side length will be $(9 \times \sqrt{2} \times 361/2)$ pm = 2 297 pm, that is, approximately 2 300 pm.

In fact, this estimate is confirmed by electron microscopy (see Section 11), which indicates that the metallic particles in the ruthenium–copper catalyst have diameters in the range 1 000 to 6 000 pm.

The information derived from an EXAFS experiment — number, type and distance from the target atom to its neighbours — is a goal of many branches of chemistry. Accordingly, EXAFS finds application in topics such as coordination chemistry, bioinorganic studies and catalysis, among others. Although EXAFS is essentially a technique for the study of bulk materials, it can be adapted in various ways to achieve surface sensitivity. In the basic EXAFS process the atom that absorbs the X-ray and emits the photoelectron is left in a hole state — that is, with a vacancy in a core orbital.

■ What subsequent changes and events might ensue from this state?

▪ As occurs in photoelectron spectroscopy (Section 5), an electron will relax from a higher level into the core level, often with the emission of an Auger electron (Section 6).

As the Auger emission is a consequence of X-ray absorption, monitoring the Auger intensity is therefore equivalent to measuring the absorption of the X-rays. Therefore, if this is done while scanning the wavelength of the incident X-rays, the result is an EXAFS-type spectrum containing fine structure, which can be interpreted in the same way as the basic EXAFS experiment. There is, however, one important distinction in application.

■ What property distinguishes X-rays and electron beams in their spectroscopic application?

▪ X-rays penetrate deeply in solid materials, whereas electrons have short path lengths in solids.

It follows that the Auger electrons actually detected in this type of experiment must have originated within the surface layers, even though the incident X-rays generate such electrons *throughout* the solid. This modification of the EXAFS experiment is, therefore, essentially a surface-sensitive technique and is called **surface extended X-ray absorption fine structure, SEXAFS**.

Figure 73a shows the iodine SEXAFS radial distribution function for one-third of a monolayer of iodine on a Cu(111) surface. In the region $R = 200–400$ pm, the main peak represents the spacing between an iodine atom and a copper atom in its nearest coordination shell, whereas the next two smaller peaks are due to the second and third coordination shells. Comparison with the corresponding EXAFS function for a bulk specimen of CuI, in which the environment of iodine is known, suggests that the adsorbed species occupies a threefold hollow site on the surface (Figure 73b). From the data in Figure 73a it is also deduced that the I(ad)–Cu(surface) bond length is 266 pm.

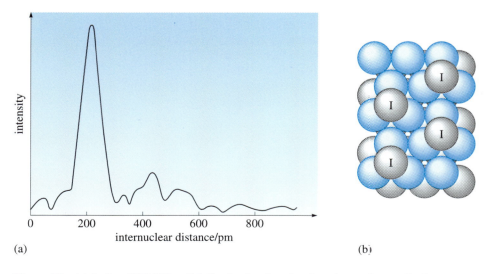

Figure 73 (a) Iodine SEXAFS radial distribution function for a ⅓ monolayer of iodine on a Cu(111) surface; (b) plan view of iodine atoms adsorbed at threefold hollow sites on Cu(111).

10.2 Summary of Section 10

1   X-ray absorption spectroscopy may be performed with a synchrotron, a tunable X-ray source.

2   An X-ray absorption spectrum exhibits extended fine structure (EXAFS).

3   An EXAFS spectrum may be converted to a radial distribution function, which gives the distances to atomic neighbours of the absorbing atoms.

4   Additionally, EXAFS provides information on the number and type of neighbouring atoms.

11 MICROSCOPY: IMAGING ATOMS

Throughout this Block we have described physical techniques that allow us to deduce properties of surfaces and molecules adsorbed on them: surface structure, chemical composition, bond strengths and orientation. The question arises as to whether we can actually 'see' atoms directly on a surface. In this final Section we look briefly at the results of some modern techniques that image surfaces.

Attempts to magnify objects date to the seventeenth century, when lenses were invented. Even the best simple hand lens will give a magnification of no more than about ×100, just enabling us to see particles containing about 10^5 atoms. Optical microscopes have magnification approaching ×1 000, so that particles of about 1 000 atoms in size can be seen. Beyond this magnification, visible light is no longer useful. By using electron beams, as in the electron microscope, the large improvement in magnification (up to about 10^5) allows terraces of atoms to be imaged as in Figure 74. To image single atoms, we must resort to some interaction between a fine probe and the individual atom on a surface. The first instrument to achieve this goal was the **field–ion microscope**, **FIM**, which employs a single gaseous atom as the probe. The equipment consists of a chamber containing the specimen in the form of a needle tip, a conducting fluorescent screen, and a noble gas (such as neon) at low pressure (Figure 75). A strong electric field is created by applying a potential difference of about 10^8 V between tip and screen, with the tip positive. Under the influence of this strong electric field, a gaseous neon atom colliding with the tip readily loses one of its electrons to the surface. The resulting Ne$^+$ ion then accelerates rapidly away from the positive specimen to strike and illuminate the negatively charged screen and produce an image of the surface site at which it was formed. Magnification is simply the ratio of the distance R (from tip to screen) to the radius r of the needle tip. With R typically about 50 mm and r about 50 nm, a magnification of about 10^6 is achieved.

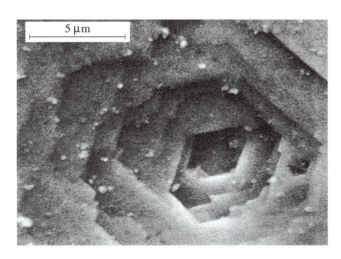

Figure 74   Electron microscope image of the surface of a zinc crystal.

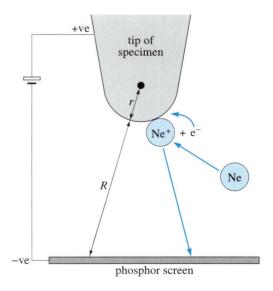

Figure 75   Schematic diagram of the field–ion microscope (tip is not to scale).

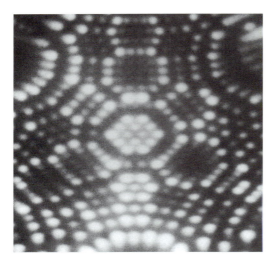

Figure 76 Field–ion micrograph of the tip of a tungsten needle.

■ Figure 76 shows an FIM image of a needle tip of tungsten, a *bcc* metal with the unit cell parameter $a_0 = 316$ pm. The bright patch at the centre of the image consists of 14 atoms in a roughly close-packed hexagonal array. Suggest which plane forms the apex of the tip and estimate the overall magnification represented by the image. (Refer to the fold-out sheet for the structure of low-index planes of a *bcc* metal.)

■ The hexagonal pattern suggests that the plane is W(111). The nearest-neighbour spacing in this plane is $\sqrt{3}a_0/2 = 274$ pm, or about 3×10^{-10} m. The spots in the image are separated by a few mm (about 3×10^{-3} m). So the magnification shown here is of the order $3 \times 10^{-3}/3 \times 10^{-10} = 10^7$.

Of course, the imaging of needle tips has limited use, but in 1982 a new instrument was developed at IBM in Zurich, the **scanning tunneling microscope**, **STM**. In the STM, a sharp probe, produced by cutting a wire so that its tip consists of just one or two atoms, is held about 1 nm above the sample surface (Figure 77). At this range, the electron cloud on a surface atom overlaps with that of an atom on the probe tip. So when a small electrical potential is applied, an electron from the surface atom appears in the probe tip — a process called tunnelling; in other words, a current flows between the tip and the surface. The exact magnitude of the current is very sensitive to the size of the gap, changing by a factor of 10 when the gap changes by 100 pm. Therefore, if the tip is scanned mechanically across a surface, the variations in current reflect the surface contours with atomic resolution. A $2\,500 \times 2\,500$ pm section of the (111) face of a crystal of gold is shown in Figure 78.

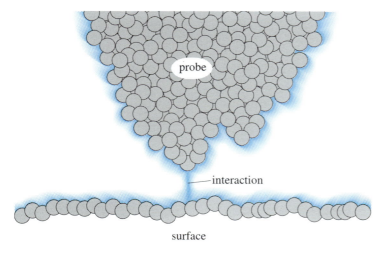

Figure 77 The probe–surface interaction

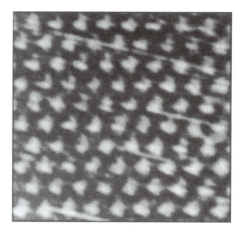

Figure 78 The STM image of the (111) face of a gold crystal, showing an area of about $2\,500$ pm across.

■ Gold has the *fcc* structure. Does the image in Figure 78 conform to your expectation of the (111) surface of such a structure?

▪ The hexagonal pattern of a *fcc*(111) surface appears to be reproduced in the STM image.

STM may be applied to many kinds of solids. The covers of the books for this Course, for example, show the image of a section of DNA, where the rows of orange–yellow peaks represent the coils of the double helix. The average distance between each peak is 3.5 nm, corresponding to the pitch of the DNA molecule (the distance between equivalent points on a helix).

Adsorbates too may be imaged, as Figure 79 illustrates. Here the copper phthalocyanin molecule is adsorbed on a Cu(100) surface. The sketch in Figure 79a shows the location of the molecule on the surface. Recall that the STM image depends on an electron being transferred from the surface to the needle tip. From a knowledge of the occupied molecular orbital of highest energy in copper phthalocyanin, it is possible to calculate the electron density in a plane located 100 pm above the surface of the molecule (Figure 79b). Notice how closely this calculation corresponds to the image generated by STM in Figure 79c.

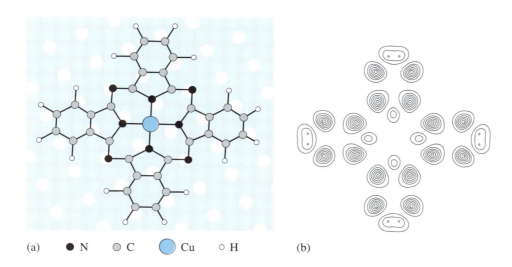

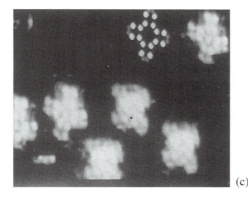

Figure 79 STM study of copper phthalocyanin adsorbed on a Cu(100) surface: (a) model of a molecule on the surface; (b) electron density in a plane about 100 pm above the surface; (c) STM image, showing relationship to (b); the dotted structure at the top shows a representation of part (b) to the same scale.

Impressive as the STM results are, an inherent weakness in the technique is its general lack of sensitivity to the chemical identity of the atoms being imaged. Although of much inferior resolution, it is possible to 'map' a surface, element by element, using a variation of a technique introduced in Section 6, Auger electron spectroscopy. An electron beam can be focused extremely sharply, to a spot of about 10^{-7} m in diameter. It is therefore possible to scan such a spot across a surface while simultaneously collecting the emitted Auger electrons from a chosen element. Not surprisingly, the technique is called **scanning Auger electron spectroscopy**, **SAES**.

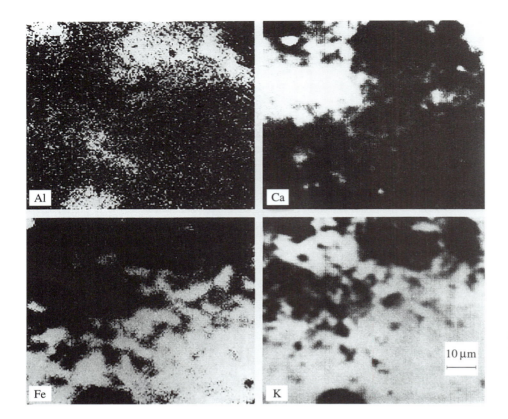

Figure 80 Scanning Auger electron spectroscopy images of the distribution of four elements, aluminium, calcium, iron and potassium, on the ammonia synthesis catalyst.

In a particularly fruitful and comprehensive study (discussed in band 6 on videocassette 2), an elemental picture has been established of the catalyst used in the production of ammonia. This catalyst consists of iron, with aluminium, potassium and calcium added as promoters. The results of the SAES study are shown in Figure 80, in which the light areas indicate the presence of the element being studied.

■ What can you conclude about the distribution and relationship of the four elements?

■ It is clear that iron and potassium occupy the same parts of the surface. On the other hand, the aluminium and calcium sites are unrelated, occurring mainly where iron and potassium are absent, although a little aluminium is scattered throughout the iron regions.

The presence of calcium and aluminium at the boundaries of the iron regions implies that these two elements act as *structural* promoters (Block 5), and prevent sintering of the iron crystallites. The scattering of aluminium in the iron region further suggests that it breaks up the iron region into tiny microcrystallites. By contrast, the intimate association of potassium and iron suggests a very different role for potassium, as a *chemical* or electronic promoter. These results confirm conclusions from other experiments, that aluminium and calcium are structural promoters, whereas potassium plays a chemical role in the catalysis.

11.1 Summary of Section 11

1 Various modern techniques exist for imaging surfaces at high magnification.

2 In the field–ion microscope, FIM, gas-phase ions formed at surface sites are projected on to a screen to generate an image with a magnification of about $\times 10^6$.

3 In scanning tunnelling microscopy, STM, a sharp probe scans a surface. Electrons are transferred from the surface to the probe to yield an image with atomic resolution.

4 By using a fine electron beam, Auger electron spectroscopy may be applied in a scanning mode (SAES), to give an elemental map of a surface.

STUDY COMMENT   A wide range of techniques has been introduced in this Block. Many of them are complementary, providing diverse results and revealing different properties. The following SAQ provides an opportunity for you to summarize the information provided by each technique, and to check that you have noted its strengths and limitations. Make sure you attempt this SAQ.

SAQ 20   For each of the techniques denoted by their acronyms below, you should now list (a) the information that is obtained, (b) the advantages and (c) the drawbacks.

1. XPS
2. UPS
3. AES
4. LEED
5. IRAS
6. EELS
7. EXAFS
8. SEXAFS
9. STM
10. SAES

APPENDIX 1 ELECTRON BINDING ENERGIES*

	K 1s	L_1 2s	L_2 $2p_{1/2}$	L_3 $2p_{3/2}$	M_1 3s	M_2 $3p_{1/2}$	M_3 $3p_{3/2}$	M_4 $3d_{3/2}$	M_5 $3d_{5/2}$	N_1 4s	N_2 $4p_{1/2}$	N_3 $4p_{3/2}$	N_4 $4d_{3/2}$	N_5 $4d_{5/2}$	N_6 $4f_{5/2}$	N_7 $4f_{7/2}$
1 H	14															
2 He	25															
3 Li	55															
4 Be	111															
5 B	188															
6 C	284															
7 N	399															
8 O	532	24														
9 F	686	31														
10 Ne	867	45														
11 Na	1 072	63		31												
12 Mg	1 305	89		52												
13 Al	1 560	118	74	73												
14 Si	1 839	149	100	99												
15 P	2 149	189	136	135												
16 S	2 472	229	165	164												
17 Cl	2 823	270	202	200	18											
18 Ar	3 203	320	247	245	25											
19 K		377	297	294	34		18									
20 Ca		438	350	347	44		26									
21 Sc		500	407	402	54		32									
22 Ti		564	461	455	59		34									
23 V		628	520	513	66		38									
24 Cr		695	584	575	74		43									
25 Mn		769	652	641	84		49									
26 Fe		846	723	710	95		56									
27 Co		926	794	779	101		60									
28 Ni		1 008	872	855	112		68									
29 Cu		1 096	951	931	120		74									
30 Zn		1 194	1 044	1 021	137		87									
31 Ga		1 298	1 143	1 116	158	107	103	18								
32 Ge		1 413	1 249	1 217	181	129	122	29								
33 As		1 527	1 359	1 323	204	147	141	41								
34 Se		1 654	1 476	1 436	232	168	162	57								
35 Br		1 782	1 596	1 550	257	189	182	70	69	27						
36 Kr		1 921	1 727	1 675	289	223	214	89		24						
37 Rb					322	248	239	112	111	30						
38 Sr					358	280	269	135	133	38	20					
39 Y					395	313	301	160	158	46	26					
40 Zr					431	345	331	183	180	52	29					
41 Nb					469	379	363	208	205	58	34					
42 Mo					505	410	393	230	227	62	35					
43 Tc					544	445	425	257	253	68	39					
44 Ru					585	483	461	284	279	75	43					
45 Rh					627	521	496	312	307	81	48					
46 Pd					670	559	531	340	335	86	51					
47 Ag					717	602	571	373	367	95	62	56				
48 Cd					770	651	617	411	404	108	67					

* All values are in electronvolts.

	K 1s	L$_1$ 2s	L$_2$ 2p$_{1/2}$	L$_3$ 2p$_{3/2}$	M$_1$ 3s	M$_2$ 3p$_{1/2}$	M$_3$ 3p$_{3/2}$	M$_4$ 3d$_{3/2}$	M$_5$ 3d$_{5/2}$	N$_1$ 4s	N$_2$ 4p$_{1/2}$	N$_3$ 4p$_{3/2}$	N$_4$ 4d$_{3/2}$	N$_5$ 4d$_{5/2}$	N$_6$ 4f$_{5/2}$	N$_7$ 4f$_{7/2}$
49 In					826	702	664	451	443	122	77					
50 Sn					884	757	715	494	485	137	89					
51 Sb					944	812	766	537	528	152	99		32			
52 Te					1 006	870	819	582	572	168	110		40			
53 I					1 072	931	875	631	620	186	123		50			
54 Xe					1 145	999	937	685	672	208	147		63			
55 Cs					1 217	1 065	998	740	726	231	172	162	79	77		
56 Ba					1 293	1 137	1 063	796	781	253	192	180	93	90		
57 La					1 362	1 205	1 124	849	832	271	206	192	99			
58 Ce					1 435	1 273	1 186	902	884	290	224	203	111			
59 Pr					1 511	1 338	1 243	951	931	305	237	218	114			
60 Nd					1 576	1 403	1 298	1 000	978	316	244	225	118			
61 Pm					1 650	1 472	1 357	1 052	1 027	331	255	237	121			
62 Sm					1 724	1 542	1 421	1 107	1 081	347	267	249	130			
63 Eu					1 800	1 614	1 481	1 161	1 131	360	284	257	134			
64 Gd					1 881	1 689	1 544	1 218	1 186	376	289	271	141			
65 Tb					1 968	1 768	1 612	1 276	1 242	398	311	286	148			
66 Dy					2 047	1 842	1 676	1 332	1 295	416	332	293	154			
67 Ho					2 128	1 923	1 741	1 391	1 351	436	343	306	161			
68 Er					2 207	2 006	1 812	1 453	1 409	449	366	320	177	168		
69 Tm					2 307	2 090	1 885	1 515	1 468	472	386	337	180			
70 Yb					2 397	2 172	1 949	1 576	1 527	487	396	343	197	184		
71 Lu					2 491	2 264	2 024	1 640	1 589	506	410	359	205	195		
72 Hf					2 601	2 365	2 108	1 716	1 662	538	437	380	224	214	19	18
73 Ta					2 708	2 469	2 194	1 793	1 735	566	465	405	242	230	27	25
74 W					2 820	2 575	2 281	1 872	1 810	595	492	426	259	246	37	34
75 Re					2 932	2 682	2 367	1 949	1 883	625	518	445	274	260	47	45
76 Os					3 049	2 792	2 458	2 031	1 960	655	547	469	290	273	52	50
77 Ir					3 174	2 909	2 551	2 116	2 041	690	577	495	312	295	63	60
78 Pt					3 298	3 027	2 646	2 202	2 121	724	608	519	331	314	74	70
79 Au					3 425	3 150	2 743	2 291	2 206	759	644	546	352	334	87	83
80 Hg					3 562	3 279	2 847	2 385	2 295	800	677	571	379	360	103	99
81 Tl					3 704	3 416	2 957	2 485	2 390	846	722	609	407	386	122	118
82 Pb					3 851	3 554	3 067	2 586	2 484	894	764	645	435	413	143	138
83 Bi					3 999	3 697	3 177	2 688	2 580	939	806	679	464	440	163	158
84 Po					4 149	3 854	3 302	2 798	2 683	995	851	705	500	473	184	
85 At					4 317	4 008	3 426	2 909	2 787	1 042	886	740	533	507	210	
86 Rn										1 097	929	768	567	541	238	
87 Fr										1 153	980	810	603	577	268	
88 Ra										1 208	1 058	879	636	603	299	
89 Ac										1 269	1 080	890	675	639	319	
90 Th										1 330	1 168	968	714	677	344	335
91 Pa										1 387	1 224	1 007	743	708	371	360
92 U										1 442	1 273	1 048	780	738	392	381
93 Np										1 501	1 328	1 087	817	775	415	404
94 Pu										1 558	1 377	1 120	849	801	422	
95 Am										11617	1 412	1 136	879	828	440	

OBJECTIVES FOR BLOCK 6

1 Recognize valid definitions of, and use in a correct context, the terms, concepts and principles printed in bold type in the text and collected in the following Table.

List of scientific terms, concepts and principles used in Block 6

Term	Page No.	Term	Page No.
adsorbate mesh, $(m \times n)$ notation	33	open surface	31
Auger effect	21	photoelectric effect	10
Auger electron	21	photoelectron spectroscopy, PES	11
Auger electron spectroscopy, AES	21	primary electrons (in AES)	22
body-centred cubic (*bcc*) structure	25	radial distribution function	55
binding energy (of electron), E_B	11	reciprocal mesh vectors, a^* and b^*	42
chemical shift effect	17	reciprocal unit mesh	40
coordination shell	56	rough surface	31
cubic close packed (*ccp*)	25	scanning Auger electron spectroscopy, SAES	61
Einstein equation	11	scanning tunneling microscopy, STM	60
electron energy loss spectroscopy, EELS	49	secondary electrons (in AES)	22
electronic band structure (of solids)	9	semiconductor	9
extended X-ray absorption fine structure, EXAFS	54	spin–orbit coupling	16
face-centred cubic (*fcc*) structure	25	step	5
far infrared	50	substrate structure	33
Fermi level, E_F	10	surface density	28
field–ion microscopy, FIM	59	surface extended X-ray absorption fine structure, SEXAFS	58
fractional surface coverage, θ	36	surface unit mesh	28
hexagonal close packed (*hcp*)	25	synchrotron	53
high-resolution electron energy loss spectroscopy, HREELS	49	terrace	5
infrared reflection absorption spectroscopy, IRAS	48	ultraviolet photoelectron spectroscopy, UPS	13
insulator	9	unit cell parameter, a_0	26
Koopmans' theorem	12	unit mesh vectors, a and b	28
LEED pattern	40	vacuum level, E_{vac}	10
low-energy electron diffraction, LEED	39	vibrational spectroscopy	47
low index planes	31	work function, ϕ	10
metal	9	X-ray absorption spectroscopy	53
Miller indices, (*hkl*) notation	26	X-ray photoelectron spectroscopy, XPS	13

2 Describe briefly, explain the principles of, and list the information about surfaces that can be obtained from the following techniques (SAQ 20):

- X-ray photoelectron spectroscopy (XPS);
- ultraviolet photoelectron spectroscopy (UPS);
- Auger electron spectroscopy (AES);
- low-energy electron diffraction (LEED);
- infrared reflection absorption spectroscopy (IRAS);
- electron energy loss spectroscopy (EELS);
- extended X-ray absorption fine structure (EXAFS);
- scanning tunnelling microscopy (STM).

3   State the relative advantages and limitations of the various techniques listed in Objective 2 for given problems in surface science. (SAQs 7 and 17)

4   Sketch energy-level diagrams illustrating the band structure of metals, semiconductors and insulators, and describe the effect of temperature on the related electrical properties. (SAQ 1)

5   Predict and interpret the relationship between the change in work function, $\Delta\phi$, and the relative electronegativities of an adsorbate and a substrate. (SAQs 2 and 3)

6   Interpret given photoelectron spectra by identifying the elements present, and recognizing the effects of chemical shift and spin–orbit coupling. (SAQs 4–6)

7   Interpret given Auger spectra by identifying the elements that are present. (SAQ 8)

8   For a *fcc* or *bcc* crystal, recognize the notation that is used to specify:

(a) a particular crystal plane, via its Miller indices, (*hkl*);

(b) a set of equivalent planes, {*hkl*};

(c) a particular direction, [*uvw*], running through the crystal lattice.
(SAQ 9)

9   Given a diagram of the atomic arrangement on a specified *fcc* or *bcc* metal surface:

(a) sketch the surface unit mesh, and label the unit mesh vectors, *a* and *b*;

(b) determine the magnitudes (*a* and *b*) of the unit mesh vectors (given their relationship to the three-dimensional unit cell parameter a_0);

(c) determine the surface density of the metal atoms;

(d) establish the directions and magnitudes of the reciprocal mesh vectors (*a*** and *b***), and hence sketch the corresponding LEED pattern.
(SAQs 10 and 11)

10   For a given adsorbate structure on a *fcc* or *bcc* substrate surface:

(a) sketch and label (using the $(m \times n)$ notation) the adsorbate mesh;

(b) calculate the fractional surface coverage of the adsorbate;

(c) use the procedure in Objective 9(d) to sketch the corresponding LEED pattern.
(SAQs 12, 13 and 15)

11   By working backwards through the procedure in Objective 9(d) or Objective 10(c), relate experimental LEED patterns to the surface structures that produced them. In particular:

(a) deduce the structure of a clean *fcc* or *bcc* surface from its LEED pattern;

(b) deduce the adsorbate structure on a *fcc* or *bcc* substrate surface from its LEED pattern;

(c) give a simple interpretation of the change in a LEED pattern caused by adsorption;

(d) comment on the significance of the intensities of LEED pattern spots.
(SAQs 14 and 16)

12   Assign features in a given IRAS or EELS spectrum to molecular vibrations. (SAQs 18 and 19)

SAQ ANSWERS AND COMMENTS

SAQ 1 (Objectives 1 and 4)

As the temperature is increased, electrons will be promoted to higher energy levels, if these are accessible. In a semiconductor the band gap between valence and conduction bands is small. So electrons will be excited from the valence band to the conduction band. Electrical conductivity depends on the presence of electrons in the conduction band. As the occupation of the conduction band increases with an increase in temperature, the electrical conductivity also increases.

SAQ 2 (Objectives 1 and 5)

Figure 7 shows that the change in work function, $\Delta\phi$, becomes more negative as the fractional coverage increases. We therefore deduce that charge is transferred *from* the adsorbed lithium *to* the tungsten substrate, thereby reducing the work function. The difference in Pauling electronegativity, $\chi(\text{Li}) < \chi(\text{W})$, leads us to conclude that the direction of charge transfer (lithium to tungsten) is consistent with the data.

SAQ 3 (Objectives 1 and 5)

An increase in ϕ corresponds to the transfer of negative charge *from* the surface *to* the adsorbate, CO. (When carbon monoxide acts as a ligand to metal atoms (via the carbon atom) in coordination complexes, electron transfer is also from metal to carbon.) The value of ϕ normally increases (or decreases) steadily towards some limiting value, as in Figures 6 and 7. In this case the behaviour is as expected until a fractional coverage of about 0.3 (as the plotted line in Figure 8 shows), when ϕ begins to increase more steeply with coverage. This behaviour has been ascribed to a change in bonding of CO to the surface that occurs with increased coverage, from top-site adsorption (where the CO is attached to one metal atom, Figure 81a) to a bridged mode (Figure 81b). There is ample evidence for this view from other techniques.

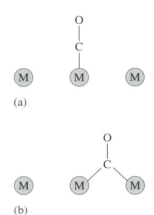

Figure 81 The adsorption of CO: (a) on top sites; (b) at 'bridged' sites.

SAQ 4 (Objectives 1 and 6)

As the photoemission is from a 3d orbital, in which spin–orbit splitting occurs, we expect the spectrum to be a doublet due to $3d_{3/2}$ (higher binding energy) and $3d_{5/2}$ (lower binding energy) components. The observation that each of these components is further split into two signals indicates that rhodium exists in two chemical environments and/or two oxidation states. The change in relative intensities with change in activity further suggests that the larger splitting is due to spin–orbit coupling, since the relative intensities of the 3/2 and 5/2 components are inherent to rhodium and should be independent of the chemical environment.

SAQ 5 (Objectives 1 and 6)

The kinetic energies quoted must first be converted to binding energies using equation 1. Thus, the peak observed with kinetic energy 1 424 eV has binding energy = 1 487 − 1 424 = 63 eV. The binding energies can then be compared with values in Table 1. When this is done, you will have noticed that the assignment of a particular peak and the identification of its source (element) is not always unambiguous. This problem is greater for large values of binding energy than for small values, as you no doubt discovered. For this reason, it is usual to look for several identifying binding energies, and to use other chemical knowledge when assigning peaks.

Starting with the peaks at lowest binding energy, the correct assignments are:

63 eV Na(2s) or Ni(3p)   Sodium is a ubiquitous element, and nickel (like iron) is a transition element and a common (sometimes added) impurity in steel.

Energy	Assignment	Note
112 eV	Ni(3s)	This confirms the presence of nickel.
165 eV	S(2p)	Sulfur is a common (sometimes added) impurity in steel.
229 eV	S(2s)	This confirms the presence of sulfur.
284 eV	C(1s)	Carbon is a common (sometimes added) impurity in steel.
532 eV	O(1s)	Oxygen is always present on metal surfaces.
872 eV	Ni(2p)	This confirms the presence of nickel.
1 072 eV	Na(1s)	This confirms the presence of sodium.

SAQ 6 (Objectives 1 and 6)

Two peaks are observed, the one at higher binding energy increasing with exposure to oxygen. This cannot be due to spin–orbit effects because these are not significant for a light atom such as magnesium (see Appendix 1) and the observed intensities change with the level of oxygen coverage. The peak at higher binding energy can be attributed to magnesium with a higher positive charge. As the intensity of this peak increases with increasing exposure to oxygen, it must be due to the formation of an oxide layer (MgO) in which the oxidation state of magnesium has changed from Mg^0 to Mg^{II}.

SAQ 7 (Objectives 1 and 3)

For most of these features, the comparison is clear cut. There is some difficulty in quantitative analysis by XPS and even more so in AES.

The completed table should be as follows:

	XPS	AES
element identification	✓	✓
sensitivity	✗	✓
analytical speed	✗	✓
quantitative analysis	d	✗
low surface damage	✓	✗
spatial information	✗	✓
chemical environmental effects	✓	✗
wide range of elements	✓	✓

SAQ 8 (Objectives 1 and 7)

The peaks marked A and B occur at electron energies of about 270 eV and 500 eV, respectively. (Note that all other peaks of significant intensity are due to nickel.) From Table 1 these peaks can be assigned to carbon and oxygen in the following way. The energy of an Auger peak is given by equation 5:

$$E_{Auger} = E_X - E_Y - E_Z \tag{5}$$

In the case of light elements such as those in the second Period, the core electron must be 1s, and so E_X is the binding energy of the 1s electron. It follows that E_Y and E_Z are the binding energies of the 2s and 2p subshells. From Table 1, we find that the values of E for 2p electrons are very small and can be ignored in this example. So taking the values of $(E_{1s} - E_{2s})$, we find that A is carbon $\{E_{Auger} = (284 - 0) \text{ eV}\}$ and that B is oxygen $\{E_{Auger} = (532 - 24) \text{ eV} = 508 \text{ eV}\}$.

Note that carbon and oxygen are always present on uncleaned metal surfaces due to adsorbed organic molecules and surface oxidation, respectively.

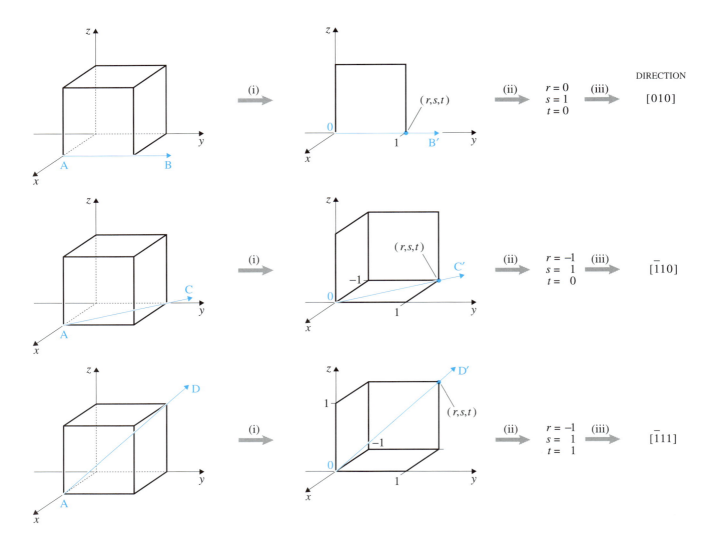

Figure 82 Steps in determining the directions in SAQ 9.

SAQ 9 (Objective 8)

The procedure you should have used to determine the directions AB, AC and AD is illustrated in Figure 82. In each case a line parallel to the direction of interest is drawn through the origin, step (i). The coordinates, r,s,t, of any convenient point on the line are then determined, step (ii). In the present examples no fractional coordinates are encountered and so r,s,t, written without commas and in square brackets, $[rst]$, can be used directly to represent the particular direction, step (iii).

SAQ 10 (Objectives 1 and 9)

From Figure 44, the bcc(110) unit mesh angle, ψ, is 109.47° and the sides are $a = b = \sqrt{3}a_0/2$. Hence the Ta(110) unit mesh area is

$$S_{Ta(110)} = (0.330 \text{ nm} \times \sqrt{3}/2)^2 \sin 109.47°$$
$$= 0.077 \text{ nm}^2$$

So the surface atom density is

$$1/S_{Ta(110)} = 12.99 \text{ nm}^{-2} = 13 \times (10^{-9} \text{ m})^{-2} = 1.3 \times 10^{19} \text{ m}^{-2}$$

SAQ 11 (Objectives 1 and 9)

The unit meshes of the three surfaces are shown in Figure 83. In the case of $bcc(532)$, the surface orientation in the question is such that the convention for representing the mesh could not be followed exactly (b horizontal).

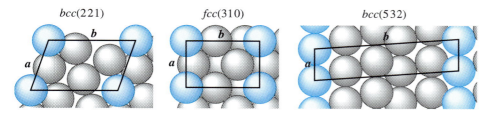

Figure 83 Answer to SAQ 11.

SAQ 12 (Objectives 1 and 10)

The unit meshes of the adsorbate structures are shown in Figure 84, outlined in black. The $(m \times n)$ notation for the adsorbed layers is as follows:

(a) $fcc(110)(2 \times 3)$ — see Figure 84a;
(b) $fcc(111)(1 \times 2)$ — see Figure 84b;
(c) $fcc(100)(\sqrt{2} \times 2\sqrt{2})R45°$ — see Figure 84c.

As shown in Figure 84c, the adsorbate mesh has a shorter side, which is equivalent to a diagonal of the (1×1) substrate mesh, and so of length $\sqrt{2}a_0$, whereas the longer side is twice this length. Also, the adsorbate is rotated by 45° relative to the substrate. Hence, the adsorbed layer is $fcc(100)(\sqrt{2} \times 2\sqrt{2})R45°$. We have seen already that on a *square* substrate mesh there is no difference, save for a 90° rotation, between (1×2) and (2×1) meshes. By considering the alternative $(2\sqrt{2} \times \sqrt{2})R45°$ structure in the present case, you might like to convince yourself that the same is also true here.

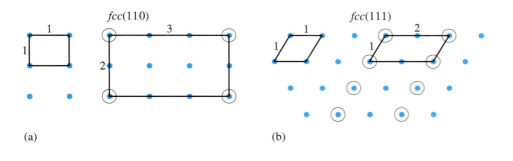

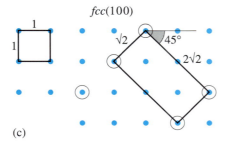

Figure 84 Answer to SAQ 12.

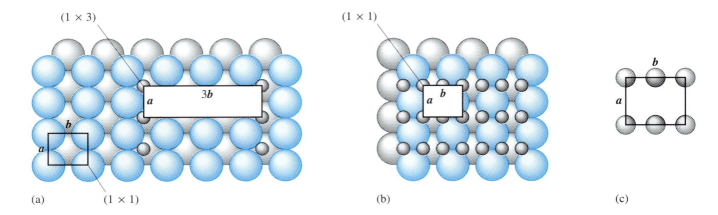

Figure 85 The labelled surface unit meshes of structures formed by hydrogen adsorbed on a Rh(110) surface: (a) at low coverage; (b) at maximum coverage. One unit mesh of the maximally covered surface is shown in (c).

SAQ 13 (Objectives 1 and 10)

Comparison of the (1×1) unit mesh of the clean Rh(110) surface with that of the low-coverage hydrogen layer shows the latter to be (1×3) (Figure 85a), so that $m = 1$ and $n = 3$. Also, there is a single hydrogen atom within each (1×3) mesh. Hence, the hydrogen fractional surface coverage, $\theta = 1/mn$, is one-third.

The mesh of the maximum-coverage structure (Figure 85b) has the same dimensions as that of the clean surface, and so is (1×1), giving $m = n = 1$. It contains, however, *two* hydrogen atoms (Figure 85c). Those at the corners, each shared between four meshes, contribute a total of one, whereas the two lying midway along the 'b' sides are shared between only two meshes, and so also contribute a total of one. Hence, the fractional coverage is $\theta = 2/1 = 2$.

For the (1×1) mesh of a clean fcc(110) surface, $a = \sqrt{2}a_0/2$ and $b = a_0$, and the mesh angle is $\psi = 90°$ (see Figure 40). Then, with $a_{0,\text{Rh}} = 0.380$ nm, the mesh area is

$$S_{\text{Rh}(110)} = (\sqrt{2}/2 \times 0.380^2) \sin 90° \text{ nm}^2 = 0.102\ 1 \times 1 \text{ nm}^2 = 0.102 \text{ nm}^2$$

Therefore, the surface hydrogen atom density at maximum fractional coverage, $\theta = 2$, is:

$$\frac{\theta}{S_{\text{Rh}(110)}} = \frac{2}{0.102 \text{ nm}^2} = 19.6 \text{ nm}^{-2} = 1.96 \times 10^{19} \text{ m}^{-2}$$

SAQ 14 (Objectives 1 and 11)

According to the examples in Figures 58 and 59, with an oblique mesh orientated in the conventional way (b horizontal and a downwards and to the left), the a^* direction in the LEED pattern will be vertically downwards, whereas the b^* direction will be downwards and to the right. The (3×1) reciprocal mesh is thus the small parallelogram shown in Figure 86 with sides of length $\frac{1}{3}a^*$ and b^*. This enables us to identify the (1×1) reciprocal mesh as the larger parallelogram, with sides a^* and b^*. Approximate measurement gives $a^*/b^* \approx 2$, and so for the real (1×1) mesh $b/a \approx 2$. The final steps are to establish the a and b directions (perpendicular to b^* and a^*, respectively) and to draw the mesh with its sides in the required ratio, as shown in Figure 86.

For information, Figure 87 shows a space-filling model of an fcc(410) surface; the precise value of b/a is $3/\sqrt{2}$ or 2.121. One possible arrangement of adsorbed species in a (3×1) layer is indicated; determination of the actual adsorption site would involve spot intensity analysis, as described in Section 8.2.3.

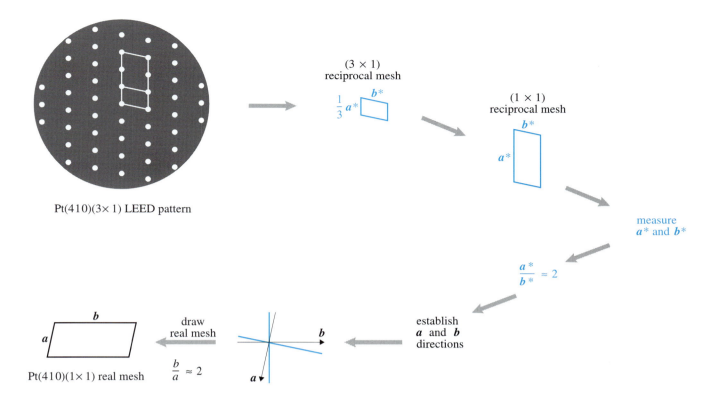

Figure 86 The steps involved in deducing the (1 × 1) unit mesh of a Pt(410) surface from the LEED pattern of a (3 × 1) adsorbate layer.

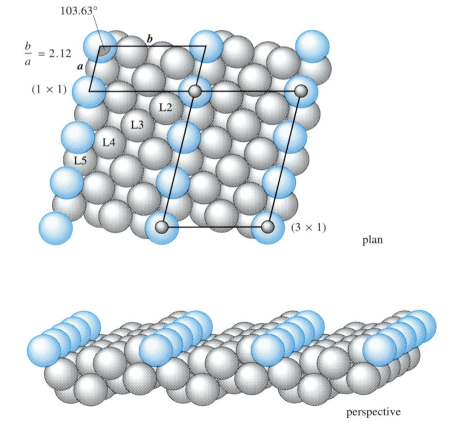

Figure 87 Plan and perspective views of a fcc(410) surface. The plan includes one possible arrangement of adsorbed species in a (3 × 1) layer.

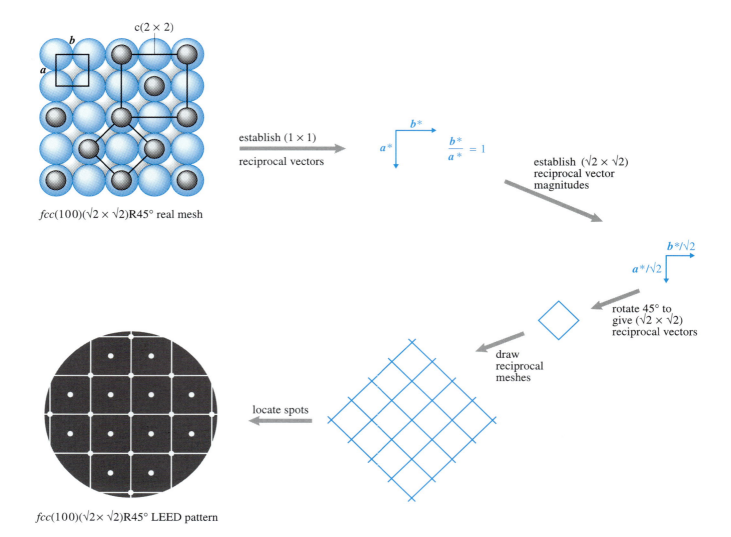

Figure 88 Steps involved in establishing the LEED pattern for a $fcc(100)(\sqrt{2} \times \sqrt{2})R45°$ adlayer structure. The reciprocal unit mesh of the clean surface is indicated by thin lines in the LEED pattern.

SAQ 15 (Objectives 1 and 10)

The $(\sqrt{2} \times \sqrt{2})R45°$ adsorbed layer is shown in Figure 88. The unit mesh of a clean $fcc(100)$ surface is square, with $a = b$, as shown at the top left of Figure 88. The adsorbed layer can be described by a centred unit mesh $c(2 \times 2)$ or by the smaller square unit mesh $(\sqrt{2} \times \sqrt{2})R45°$, both of which are shown in Figure 88.

The reciprocal unit mesh of a $fcc(100)$ surface is square, with $a^* = b^*$. So the reciprocal vectors of the $(\sqrt{2} \times \sqrt{2})R45°$ adlayer will have magnitudes $a^*/\sqrt{2} = b^*/\sqrt{2}$ and directions lying at 45° to $a^* = b^*$. The steps involved in establishing the corresponding LEED pattern are shown in Figure 88.

SAQ 16 (Objectives 1 and 11)

The pattern in Figure 62a has a square reciprocal mesh and so corresponds to a square lattice in real space. In the bcc system (Figure 44) the low-index plane with a square unit mesh is (100). (You may have recognized that Figure 62a is a repeat of the LEED pattern in Figure 54b.)

In the LEED pattern for the oxidized surface (Figure 62b) the spot spacings in both the a^* and b^* directions are reduced by a factor of about two, compared with the clean surface pattern. Hence, the oxidized structure is probably $W(100)(2 \times 2)$–O.

SAQ 17 (Objectives 1 and 3)

The comparison is contained in the following table:

IRAS	EELS
used at high pressure; relevant to catalysis	used at low pressure; not under catalytic conditions
high resolution (≈ 1 cm^{-1})	low resolution (≈ 30 cm^{-1}), except with special instruments
limited spectral range	wide spectral range
high sensitivity (10^{-3} monolayer)	lower sensitivity
selection rules limit observed vibrations	selection rule applies in specular direction

SAQ 18 (Objectives 1 and 12)

According to equation 22, this low-frequency absorption is characteristic of a vibration involving a heavy atom, and therefore a large value of reduced mass, μ. In this case the absorption could be attributed to a Rh—C or a Rh—O vibration, but without further information it would not be possible to make a more definite assignment.

SAQ 19 (Objectives 1 and 12)

At 110 K the frequency is close to that of an N=N vibration. This indicates that the nitrogen is non-dissociatively adsorbed (probably as a bridged structure with each nitrogen atom bound to neighbouring iron atoms). The low-frequency band at 170 K suggests a vibration involving a heavy atom (see equation 22). This is likely to be due to a Fe—N species. Therefore, the appearance of this band, together with the disappearance of the band at 1 490 cm^{-1}, indicates that the nitrogen adsorbed molecularly at 110 K dissociates on the surface when it is heated to 170 K.

SAQ 20 (Objectives 1 and 2)

1   XPS: X-ray photoelectron spectroscopy

(a)   Electrons from core levels are studied to determine their binding energies. These give an elemental analysis of the surface and sometimes the oxidation state of the element and information about bonding.

(b)   High sensitivity and universal applicability as an analytical method; high surface sensitivity.

(c)   Applicable only at ambient temperature and low pressure.

2   UPS: Ultraviolet photoelectron spectroscopy

(a)   Electrons from the valence and conduction bands are studied to determine their binding energies. Adsorbates are studied, sometimes giving details of their bonding to the surface.

(b)   High spectral (energy) resolution is possible and the method has quite high surface sensitivity (but less than XPS).

(c)   Applicable only at ambient temperature and low pressure.

3   AES: Auger electron spectroscopy

Similar to XPS. Advantages over XPS include greater sensitivity and speed, and the provision of spatial information. However, surface damage may occur and information about the chemical environment is limited.

4 LEED: Low-energy electron diffraction

(a) The diffraction of an electron beam by the surface gives the structure of the substrate and of an adsorbate layer if present.

(b) Direct study of surface structure.

(c) Applicable only to single crystal studies at ambient temperature and low pressure — that is, under non-catalytic conditions.

5 IRAS: Infrared reflection absorption spectroscopy

(a) Identity and structure of molecules adsorbed on surfaces and information about surface–adsorbate bonds.

(b) High spectral resolution.

(c) Selection rule limits the observed vibrations (sometimes to advantage); limited spectral range.

6 EELS: Electron energy loss spectroscopy

(a) As IRAS.

(b) Wide spectral range in single experiment.

(c) Lower resolution than in IRAS. Operates only at low pressure.

7 EXAFS: Extended X-ray absorption fine structure

(a) The absorption of X-rays reveals information about the environment of the absorbing atom: coordination, type and number of nearest neighbours, and distances to them.

(b) Detailed (unique) structural information; sample need not be crystalline.

(c) Bulk properties may dominate over surface species; synchrotron source required.

8 SEXAFS: Surface EXAFS

(a) As EXAFS.

(b) Surface sensitive.

(c) Low pressure if Auger electron emission is used; synchrotron source required.

9 STM: Scanning tunnelling microscopy

(a) Surfaces are imaged at the atomic level.

(b) Highest spatial resolution images provided; individual atoms are observed.

(c) Can operate at *atmospheric* pressure. No chemical information.

10 SAES: Scanning Auger electron spectroscopy

(a) As Auger spectroscopy with spatial mapping of each element's distribution.

(b) Unique element-imaging of surface.

(c) Operates at low pressure; resolution is much inferior to STM.

ACKNOWLEDGEMENTS

Grateful acknowledgement is made to the following sources for permission to reproduce material in this Block:

Cover photograph: Lawrence Berkeley Laboratory/Science Photo Library; *Figure 2*: G. A. Somorjai, *Introduction to Surface Chemistry and Catalysis* (1994), copyright © 1994 John Wiley and Sons Inc., reprinted by permission of John Wiley and Sons Inc; *Figures 3 and 21a and b*: D. P. Woodruff, 'Electron spectroscopies', in D. P. Woodruff and T. A. Delchar (eds), Modern T*echniques of Surface Science* (2nd edn, 1994), Cambridge University Press; *Figure 6*: reprinted from P. J. Goddard and R. M. Lambert, *Surface Science, 67* (1977), (1), 'Adsorption–desorption properties and surface structural chemistry of chlorine on copper(111) and silver(111)', p. 180, with kind permission of Elsevier Science–NL, Sara Burgerhartstraat 25, 1055 KV Amsterdam, The Netherlands; *Figure 7*: V. M. Gavrilyuk and V. K. Medvedev, 'Investigation of the adsorption of lithium on the surface of a tungsten single crystal in a field-emission projector', Physics *of the Solid State*, **8** (1966), (6), American Institute of Physics; *Figure 8*: reprinted from C. M. Mate, C. T. Kao and G. A. Somorjai, *Surface Science, 206* (1988), 'Carbon monoxide induced ordering of adsorbates on the Rh(111) crystal surface: importance of surface dipole moments', p. 145, with kind permission of Elsevier Science–NL, Sara Burgerhartstraat 25, 1055 KV Amsterdam, The Netherlands; Fi*gure 10*: *Chemistry Society Summer School in Photoelectron Spectroscopy*, University College of Swansea, 16–21 April, 1972, The Royal Society of Chemistry; *Figures 13 and 21c:* D. W. Turner *et al.*, *Molecular Photoelectron Spectroscopy* (1970), John Wiley and Sons Ltd, reprinted by permission of John Wiley and Sons Ltd; *Figure 15*: C. S. Fadley, 'Basic concepts of X-ray photoelectron spectroscopy', in C. R. Brundle and A. D. Baker (eds), Electron *Spectroscopy: Theory, Techniques and Applications*, vol. 2, Academic Press (1978) (London) Ltd; *Figure 16*: W. M. Riggs *et al.*, *Handbook of X-Ray Photoelectron Spectroscopy*, © 1979 by Perkin–Elmer Corporation, Physical Electronics Division, all rights reserved; F*igure 17*: C. Nordling, 'Electron spectroscopy for chemical analysis', Angewa*ndte Chemie*, International Edition, **11** (1972), VCH Verlagsgesellschaft mbH; Figure 18: A. D. Baker *et al.*, Chemical So*ciety Reviews*, Vol 1 (1972), The Royal Society of Chemistry; *Figure 20*: A. F. Carley *et al.*, 'Oxygen dimerization at a Zn(0001)–O surface', Journal *of Solid State Chemistry*, **112** (1994), Academic Press Inc., all rights reserved; *Figure 24*: R. Holm, 'Imaging and analysis of surfaces with a scanning electron microscope and electron spectrometer', Angewan*dte Chemie,* International Edition, **10** (1971), (9), VCH Verlagsgesellschaft mbH; Figure 25: adapted from K. Siegbahn, Electron Spectroscopy — *an Outlook*, Kai Siegbahn (1974), Uppsala University; *Figure 30*: G. Ertl, 'Elementary processes at gas/metal interfaces', Angewan*dte Chemie, International Edition*, **15** (1976), (7), VCH Verlagsgesellschaft mbH; Figure 60: reprinted from H. Ohtani, M. A. van Hove and G. A. Somorjai, 'LEED intensity analysis of the surface structures of Pd(111) and of CO adsorbed on Pd(111) in a ($\sqrt{3} \times \sqrt{3}$)R30° arrangement', *Surface Science, 187* (1987), p. 375, with kind permission of Elsevier Science–NL, Sara Burgerhartstraat 25, 1055 KV Amsterdam, The Netherlands; *Figures 64 and 65*: N. V. Richardson, 'Electrons to stimulate vibrations', Chemist*ry in Britain*, **29** (1993), (1), The Royal Society of Chemistry; *Figure 67*: M. Grunze *et al.*, 'π–bonded N2 on Fe(111): the precursor for dissociation', Physical *Review Letters*, **53** (1984), (8), The American Physical Society; *Figures 68 and 69*: S. P. Cramer and K. O. Hodgson, 'X-ray absorption spectroscopy: a new structural method and its applications to bioinorganic chemistry', *Progress in Inorganic Chemistry*, **25** (1979), John Wiley and Sons Inc., reprinted by permission of John Wiley and Sons Inc; Figure 70: E. A. V. Ebsworth, D. W. H. Rankin and S. Cradock, *Structural Methods in Inorganic Chemistry*, Blackwell Science Ltd (1987); *Figure 71b and c*: J. M. Hollas, Modern *Spectroscopy*, John Wiley and Sons Ltd (1987), reprinted by permission of John Wiley and Sons Ltd; *Figure 72*: J. H. Sinfelt, G. H. Via and F. W. Lytle, 'Structure of bimetallic clusters. Extended X-ray absorption fine structure (EXAFS) studies of Ru–Cu clusters',

Journal of Chemical Physics, **72** (1980), (9), American Institute of Physics; Figure 73a: reprinted from P. H. Citrin, 'X-ray absorption spectroscopy applied to surface structure: SEXAFS and NEXAFS', *Surface Science, 299/300* (1994), p. 210, with kind permission of Elsevier Science–NL, Sara Burgerhartstraat 25, 1055 KV Amsterdam, The Netherlands; *Figure 74*: Professor G. A. Somorjai; F*igure 76*: M. K. Miller and G. D. W. Smith, A*tom Probe Microanalysis: Principles and Applications to Materials Problems*, Materials Research Society (1989), Pittsburgh; *Figure 77*: reprinted from H. Rohrer, 'Scanning tunneling microscopy: a surface science tool and beyond', Surface Science, **299/300** (1994), p. 958, with kind permission of Elsevier Science–NL, Sara Burgerhartstraat 25, 1055 KV Amsterdam, The Netherlands; *Figure 78*: V. M. Hallmark *et al.*, 'Observation of atomic corrugation on Au(111) by scanning tunneling microscopy', Phys. *Rev. Lett,* **59** (1987), (25), p. 2 880; *Figure 79:* S. Chiang and R. J. Wilson, 'Surface imaging by scanning tunneling microscopy', in D. B. Williams, A. R. Pelton and R. Gronsky (eds), Images of *Materials*, Oxford University Press (1991), by permission of Oxford University Press; *Figure 80*: G. Ertl, *et al.*, 'Surface characterisation of ammonia synthesis catalysts', Catal*ysis Reviews*, **21** (1980), Part 2.

Journal of Chemical Physics, **72** (1980), (9), American Institute of Physics; Figure 73a: reprinted from P. H. Citrin, 'X-ray absorption spectroscopy applied to surface structure: SEXAFS and NEXAFS', *Surface Science, 299/300* (1994), p. 210, with kind permission of Elsevier Science–NL, Sara Burgerhartstraat 25, 1055 KV Amsterdam, The Netherlands; *Figure 74*: Professor G. A. Somorjai; Fi*gure 76*: M. K. Miller and G. D. W. Smith, A*tom Probe Microanalysis: Principles and Applications to Materials Problems*, Materials Research Society (1989), Pittsburgh; *Figure 77*: reprinted from H. Rohrer, 'Scanning tunneling microscopy: a surface science tool and beyond', Surface Science, **299/300** (1994), p. 958, with kind permission of Elsevier Science–NL, Sara Burgerhartstraat 25, 1055 KV Amsterdam, The Netherlands; *Figure 78*: V. M. Hallmark *et al.*, 'Observation of atomic corrugation on Au(111) by scanning tunneling microscopy', Phys. *Rev. Lett,* **59** (1987), (25), p. 2 880; *Figure 79:* S. Chiang and R. J. Wilson, 'Surface imaging by scanning tunneling microscopy', in D. B. Williams, A. R. Pelton and R. Gronsky (eds), Images of *Materials*, Oxford University Press (1991), by permission of Oxford University Press; *Figure 80*: G. Ertl, *et al.*, 'Surface characterisation of ammonia synthesis catalysts', Catal*ysis Reviews*, **21** (1980), Part 2.